PLANETFALL

books by MARTIN CAIDIN

Planetfall
The Tigers Are Burning
Destination Mars
The Fork-Tailed Devil
Zero!
Samurai!
The Ragged, Rugged Warriors
Black Thursday
Thunderbirds!
The Long Arm of America
The Long, Lonely Leap
The Mighty Hercules
Let's Go Flying!
Cross-Country Flying
Vanguard!
Countdown for Tomorrow
The Greatest Challenge
Aviation & Space Medicine
Test Pilot
By Apollo to the Moon
Hydrospace
Rendezvous in Space
Man-in-Space Dictionary
Spaceport USA
The Astronauts
The Winged Armada
War for the Moon
Air Force
Thunderbolt!
Golden Wings
This Is My Land
Everything But the Flak
A Torch to the Enemy
The Night Hamburg Died
Barnstorming
Boeing 707
Flying Fort

Flying
Rockets Beyond the Earth
I Am Eagle!
Red Star in Space
Jets, Rockets & Guided Missiles
Rockets & Missiles
Man into Space
Messerschmitt Me-109
The Zero Fighter
New World for Men
The Mission
The Silken Angels
The Long Night
Why Space?
First Flight into Space
Worlds in Space
Overture to Space
When War Comes
It's Fun to Fly
The Power of Decision

NOVELS

Cyborg
Marooned
The Last Dogfight
Operation Nuke
Devil Take All
No Man's World
The God Machine
Anytime, Anywhere
The Last Fathom
The Mendelov Conspiracy
Four Came Back
The Cape
Almost Midnight
Maryjane Tonight at Angels Twelve

PLANETFALL

by Martin Caidin

Coward, McCann & Geoghegan, Inc.
NEW YORK

Copyright © 1974 by Martin Caidin

All rights reserved. This book, or parts thereof, may not be reproduced in any form without permission in writing from the publisher. Published on the same day in Canada by Longman Canada Limited, Toronto.

SBN: 698-10536-2
Library of Congress Catalog Card Number: 73-78752
Printed in the United States of America

*In the beginning there were only
a few who believed. There were
less who worked at it. This book
is for a man who has done both
since it all started . . .*
                    GORDON L. HARRIS

Contents

1.	Careful, There	11
2.	A Wonderful, Impossible World	22
3.	Planetfall on Aeolus	32
4.	Third Planet from the Sun	39
5.	The Splendid Destiny	45
6.	Prelude	59
7.	Outbound	76
8.	Road Map of Space	89
9.	Time Machine	108
10.	Sister World	133
11.	Into the Atmospheric Deeps	146
12.	Those Incredible Robots	158
13.	The Mars We Never Knew	181
14.	Mercury to Mars—the New Robots	200
15.	The Outworld	218
16.	The Rocky Road to Space	228
17.	Steamboat Time	234
	Index	247

PLANETFALL

Chapter 1

Careful, There

Think of it. At this moment, *right now,* a strange, complex and dazzling robot is sailing through a sea of vacuum toward the planet Jupiter.

Another three robots are whirling in high, looping orbits about the planet Mars.

A whole family of robots is orbiting the Earth's moon. More robots stand on the surface of the moon, drawing electrical power from strange nuclear reactors, recording temperature, radiation, the "wind" of particles flowing from the sun and from deep space. If the moon quivers, the atomic-powered instruments record every tremble and sigh, and send the information across a quarter of a million miles of space back to waiting receivers on the Earth. Other instruments electronically "sniff" for the faintest sign of gases that may be seeping away from within the battered surface of the moon.

Other robots crawl slowly on flimsy treads through the gray dust of the moon.

A robot sits silently on the surface of Mars.

Still another robot is racing toward the most beautiful of all worlds of the solar system, the ringed wonder that is Saturn.

Several robots lie crushed beneath the massive, squeezing atmosphere of Venus.

And more are being prepared for their adventures away from Earth. Probes to orbit and then to land on Venus. Huge ships to slide into orbit about Mars and send robot laboratories to its surface. More robots to wing toward Mercury. Still more to Saturn, and then outward bound to the edges of the solar system.

A wave spreading out from Earth in a miracle of exploration. Electronic magic to link us with worlds small and large, some frozen solid, others torn with heat, some with their surface nakedly exposed to the harsh radiations of the sun, others concealed beneath atmospheres so thick their "air" flows like liquid metal.

The worlds of our solar system, the family of planets that slide endlessly through the vast reaches of space about our sun. The planets and the thousands of asteroids and planetoids and comets and meteoroids: huge bodies, others small and rocky and battered into celestial junk.

That's where our interest lies, in Mercury and Venus and Earth and Mars, in Jupiter and Saturn, in Uranus and Neptune and Pluto, in their moons and all the wondrous debris of our own backyard in space.

We're interested in all those worlds and smaller bodies "out there" as we know them at present, and how much better we're going to know them in the coming months and years. Far out in space, closer to and farther away from the sun than is our own globe, are planets mysterious and fascinating to us. Until this age, until this very time in which we live, we had no choice but to guess what those worlds really were like. Some of those guesses were wild. Not until recently, not until *now,* did we learn just how wild—how very wrong—we were.

Even the conclusions that were drawn on the basis of the most serious scientific study often turned out to be devastatingly wrong.

Until just a few years ago we stumbled along in ignorance even about our moon which, at less than a quarter million miles from us, is barely more than a small hop where space is concerned. The moon, as far back as our history records and man looked up into the night sky, fascinated and beguiled us. It was so near, and yet so far!

But at least we could study the moon. We looked at it with

the naked eye and drew pictures of its details, and continued to prepare such charts through the years so that we might note any changes on its seemingly unchangeable surface. We built telescopes and then supertelescopes to record every visible feature on photographic plates so that we might study them in leisure. We built special tools—spectroscopes, radar and other wonderful devices—to unlock the secrets of a world that lay bare and barren to us, uncloaked by not even the barest shred of murky atmosphere. We did all these things, and after many years of serious study by the most respected men of science, those same scientists still could not agree on what the moon was really like.

They couldn't agree about the craters on the moon. There they were, thousands upon thousands of them, glaringly visible, and the scientists argued with one another about the craters being formed by huge meteorites or volcanic blasts hundreds of millions of years ago. Or perhaps they were a combination of the two.

Today, Project Apollo is history. Man first landed on the moon with the slow climb down the ladder of the first spacecraft by Neil Armstrong. But only a year or two before Apollo XI began its fiery ride down to the lunar surface, a heated debate raged among top scientists about the nature of that surface. Some men, and they included some of the most brilliant men in science, were so certain the moon's surface was a deep layer of dust that they did everything possible to prevent the first manned landing on the cratered world a quarter of a million miles from Earth. They forecast a dire moment when the first spacecraft would shut off its blazing engine and then sink slowly and forever from sight in a vast sea of dust from which it would never emerge.

It may sound like fiction to us *now,* but just a few years ago it was considered a theory as valid as any other.

Think of the mountains of the moon. Today they're familiar to every human being who has access to a newspaper, a magazine, a book, a television or a motion picture screen. Incredible pictures—taken with still camera, motion picture cameras and

PLANETFALL

television cameras—in color of the gently rounded surfaces of that world hanging so brightly in our night sky.

Round, gentled, smoothly flowing slopes. The result of billions of years of searing radiations from the sun, of cosmic rays from deep space, of battering by tiny and giant particles plunging from space, of billions of years of alternating heat and cold that expanded and contracted the materials on the surface. What those mountains *looked like* should have been obvious to us. After all, there they were, right before our eyes, brought to great magnification by huge telescopes. The pictures told the story, and how could so many thousands of pictures be wrong?

We *saw* the sharp shadows. *Saw* them. Shadows that could be made only by towering mountains with jagged edges, with great sawtoothed peaks. Mountains rearing high and deadly with sharp crags and needlelike upthrusting and terrible edges. Rock so sharp that to fall or brush against these granite razorlike projections must mean certain death for a man in a pressure suit. There were those mountains and cliffs, peaks and ranges, existing in a world of vacuum. Without air, there could be no wind to grind them down. There was no water to erode their surfaces. No snow or ice or savage bolts of lightning.

We looked at the moon and we saw what it was like, and seeing is believing, and we were right about the absence of sandpapering, rasping erosive forces like air and water and wind and lightning and chemicals. We looked at those huge shadows stretching ominously across the lunar surface and we knew, for we had *proof*, that the mountains of the moon were things to be feared.

But we always wanted better pictures. Taken close up. And suddenly we had a way to get those photographs. We had powerful rockets at our disposal. Along with the rockets we had marvelous new sources of energy, small enough to be included in scientific payloads, that could power television cameras. We perfected ways to transmit TV pictures across a quarter-million miles of space. We would send instruments and TV cameras to the moon, and we would get remarkable close-up photographs of that mysterious but visible surface.

Easier said than done. The first time we tried was on August 17, 1958, with a rocket called the Thor-Able and a small payload named Pioneer. I was present at the launching of every lunar probe; it promised to be a long list of probes, because we really didn't do very well in getting those expensive and wonderful payloads from the launching pads to the moon. Pioneer I made it off the pad in a beautiful launching at dawn, but 77 seconds after its blazing lift-off, when it was 50,000 feet high, the rocket exploded. The second Pioneer made it away from the Earth to a distance about one-third of the way to the moon, and then fell back toward the Earth. The third mission got off to a beautiful start, and then a wire—that's all, just one wire—snapped, and the whole thing fell back toward the Atlantic Ocean and smashed into the sea. Those were all with Thor-Able rocket boosters.

Well, we tried with two Juno II boosters. We managed to get one of them up to escape velocity, and it went sailing out into space, but it missed the moon by 37,000 miles, and really didn't do anyone much good. Then came the series with Atlas-Able boosters, huge rockets of advanced design and enormous power. The first one of those caught fire during a ground test, blazed merrily, fell over and exploded. The remaining three all failed for one reason or another.

Try and try again. We went to the series known as Atlas-Agena (which ultimately proved to be a winner). The first six Atlas-Agena shots to the moon continued the embarrassment of failures. Invariably the rocket boosters worked fine, but minute or minor difficulties in the complex systems plagued the program. No more Pioneers for a while to the moon; the new series was known as Ranger. For a while they thought of renaming the program Frustration. The Ranger spacecraft, their instruments silent and useless because a thrown switch or broken wire drained away all electrical power, one after the other tumbled like mindless idiots toward the moon. Some barely scraped by the high mountains of the moon, others whirled away uselessly into orbits about the sun, still others crashed at high speed into the lunar surface. Ranger VI attained a new height of irony. The Atlas-Agena rifled the complex Ranger with all its instruments and cameras into what was a perfect slide across space to the moon. The climax to this brilliant rocket-powered slingshot of 239,000 miles came as Ranger VI zeroed in to its target point—and all systems went utterly dead as the power systems failed.

Ranger VII cracked open the sky with thunder on July 28, 1964, and broke the jinx, sending back 4,308 TV pictures before it destroyed itself after a beautiful flight. Ranger VIII, on February 17, 1965, repeated the success with 7,137 photos. Then, on March 21, 1965, Ranger IX went aloft on a pillar of fire, buoyed as much by prayers and curses as by rocket thrust. Navigation and control were dead-on, and IX sailed moonward on schedule, directly toward the center of the huge crater Alphonsus and its jagged mountains, near the lunar equator. The cameras were working beautifully, aimed at the exact point where it would crash at thousands of miles an hour. But up until the very last instant of flight, before it would explode to create a small crater of its own to mark its passing, it would transmit back to Earth the long-sought close-up photos of stark mountains on the cratered floor of Alphonsus, and of the jagged walls that made up the high rim of the crater.

The pictures shocked almost everyone in science. Where were those terrible mountains? The jagged peaks? Instead of what we *expected* to see, the mountains in the center of Alphonsus were almost friendly. They were so round that many of them could almost be regarded as gentle hills—which is precisely what many of the lunar mountains turned out to be.

Those of us who had spent years in writing of the moon and what it would be like on that cinderous globe sighed, reached into our reference files, and started throwing away reams of material telling us what the moon certainly would be like. We had, like the scientists, looked at the photographs and, like the scientists, we had all forgotten the painful lesson that a shadow is not always a true representation of an object. The lunar shadows, tremendously long and terribly stark in that world of vacuum, were far more intense than we had thought (because we really didn't think hard enough or long enough about it), and they misled us completely.

Since that eye-opening mission of Ranger IX a small army of robots has crossed the radiation void between Earth and the moon. With every new flight we gained clearer pictures and increased our expanding book of knowledge on the characteristics of the moon. Pioneer, Ranger, Orbiter, Lunik, Luna, Surveyor, IMP, and, finally, Apollo.

Men walked the moon. As far as the lunar surface was concerned, even in the highest and most tumbled of all mountains, it was a fairly gentle world. True, it didn't have any atmosphere, and if you ran out of oxygen in your spacesuit you didn't have much time to say your last message. But there are rules to follow for every place, and if you followed them on the moon—as did our astronauts—then there were compensations.

No wild animals. No sudden raging storms. No fog to blind you, no snow or ice underfoot. Nothing to attack from the sky. To be sure, the moon has moonquakes. Some of them are rather severe. It also is vulnerable to pieces of rock tearing out of the vacuum sky at anywhere from five to a hundred thousand miles an hour; being hit with something like that can ruin your whole day. But the troublesome quakes are so rare, as are the big meteoroids zinging in from somewhere "out there," that the odds of being done in by such dangers are so remote as to be ignored. A man who was once struck by lightning in Minnesota really doesn't have to look over his shoulder for fear the same thing is likely to happen to him in the Sahara Desert.

All of which brings us to a painful lesson.

Few things are what they seem to be, most especially to the naked eye.

PLANETFALL

A second lesson in this classroom of our solar system is that the camera, when you get down to it, is really a dumb creature. All it can do is to record the reflection of light from a surface, or the origin of light from something like a fire, a flashlight, or a star. It doesn't record what's actually there. Like the human eye it is a dangerous messenger unless we understand just about all the circumstances surrounding any situation where we use a camera.

The photographs taken of the jagged shadows on the surface of the moon are an excellent example. Because the shadows were jagged (and they still are), it didn't mean that the source of those shadows also was jagged. It just *looked* that way, to the human eye and the camera. We were fooled, or honestly mistaken. Whatever the degree of blame we were *wrong*.

Today when we study pictures of the moon we *know* the lunar mountains are wonderfully gentle, with rounded peaks and domes, and they look (in physical outline) like just what we would expect to find on Earth in an area we might describe as "rolling hills." The point is that we now have experience with this sort of thing, and when we look at photographs of stark shadows we *know* the mountains from which those shadows stretch aren't jagged. So we know what to expect.

Much the same problem exists with certain pictures of craters on the moon. The only time to get excellent crater photography is when the angle of light striking the craters is low, what we would call a low side angle. This increases dramatically the lights and shadows, or the starkness of contrast, of the picture, and details leap forward vividly.

However, you've got to be very careful here. If you hold your photograph one way the craters are *not* craters—they become mounds or round domes protruding upward from the surface. That's what they look like and how the eye interprets what it sees. Turn the picture to one side, or upside down, and the mounds vanish, to become craters.

This happens even when the viewer *knows* his eye can fool

him. But having experience with the problem, recognizing the danger, he knows enough to keep turning that picture until he sees craters—no matter what his eye insists upon telling him about mounds.

The point to all this is that we're making explorations of the moon and the planets and *their* moons with scientific probes that use many wonderful instruments, counting among their number extraordinary television cameras that transmit pictures across tens of millions of miles. Those cameras are a modern sorcerer's dream come true. If ever we could be witness to magic and miracle wrapped up in one gleaming package, then the camera systems (which include transmission, receiving, computer enhancement and print-out) have to be placed at the head of the list.

Yet we must be careful how we interpret not what we see but what the camera has recorded. And one of the best lessons in interpreting photography taken at different points about the solar system is found right here on Earth, with a situation as old as man himself.

The mirage.

Careful, there. Don't leap to conclusions. The odds are the mirage is *not* what you think it is.

First of all, the mirage—to the human eye and *to the camera* —is real. In optical terms (and this includes our vision, remember), the mirage is as real as anything else we may see in our lives.

It is not hallucination, but you'd be surprised at how many people lump the two together. Hallucination remains strictly within the mind. Because of some type of mental stress, the mind creates a picture you think you're seeing—but if you're the victim of hallucination you're the only one who can see what's going on before you. It's completely self-inclusive, a private showing all your own.

Not so the mirage. If you have a hundred people together, then every one of those people will see the mirage. Let's say you're walking in the desert. You look up at the sky and there—vivid, astonishing, utterly real, and utterly impossible—you see a sailboat gliding majestically through the air.

You *know* this is impossible. You shake your head and rub your eyes and likely you even pinch yourself, but what gets you even more confused is that any companions with you also see that sailboat moving briskly before a cooling sea breeze.

You grab for your camera, and in this case we'll say it's a Polaroid, so we can check out what the camera sees, or doesn't see, within sixty seconds of snapping the picture. And when you take that picture and develop it within sixty seconds you're in for a shock—because the sailboat and the blue water and those puffy clouds that simply cannot be there *are* there, recorded permanently on the film.

What now? Obviously while you can hallucinate, your camera cannot. But you *know* the sailboat isn't there! Right. Then what are you seeing?

Reflected light, but in this case, light that may be reflected from the surface of a lake or a river hundreds of miles away. The atmosphere contains some pretty cute tricks for all of us. One of those tricks is that air isn't as simple as it seems to be. It's filled with layers of warm air and more layers of cold air and what we also call inversion layers. Sometimes these perform in the manner of mirrors or invisible screens in the sky. Sunlight striking a lake at an angle is reflected at an opposite angle back into the sky. Normally we don't see this process of reflection.

But if there's an inversion layer between the observer (in this case, you) and the distant lake, that reflection is "played with full-screen clarity" on that layer of air, where it becomes dramatically visible. You see it, and your camera sees it, and if you didn't know the circumstances under which that picture was taken, there would be no way to understand that you were taking a picture of a reflection in the sky—a mirage.

(Without such reflection we couldn't see anything about us. If an object is perfectly black it can't reflect any light. Being perfectly black it absorbs all radiations that strike its surface and, since it doesn't reflect any radiations, it doesn't reflect visible light. It remains invisible to the human eye *and* the camera.)

Exploring the moon by camera, as we have seen, meant tripping over many old and cherished beliefs. Something like that is enough to shake you up for a long time, but you'd be surprised how many men—and this most especially includes men of science—can be hoodwinked into making the same sort of mistake twice. This actually happened when our Mariner planetary probes sent back to Earth the first dramatic, marvelous photographs of the planet Mars.

We examined those truly epochal pictures. For the first time we were looking at close-up photography of a distant world, another planet. What we saw was clear enough. Craters. Huge craters marching along the Martian surface.

A man seeking information draws that information from any and all sources. The photographs of Mars were the most dazzling lodestone of planetary investigation we had ever known in our history. There, in one stroke, the burdensome mysteries of what might be the surface of Mars were banished. For now we had photographic proof of what Mars was like, and, for the first time, we were able to draw conclusions based upon optical-photographic studies of the distant, red planet.

And we were—once again!—devastatingly wrong.

But that's another chapter.

Chapter 2

A Wonderful, Impossible World

Have you ever heard of the planet Aeolus?
No?
Well, that's hardly surprising, since Aeolus is a fictional world created by scientists. A world teeming so richly with different forms of life that it would be utter paradise for a biologist. Aeolus is an exercise in imagination for serious reasons. By creating such a world, and specifying the conditions that exist on the planet, even with the seas that we accept will be there, we can carefully work out forms of life that otherwise might lie beyond our wildest dreams for the planet we call Earth. Being able to extrapolate in this fashion enables scientists to pursue a "What if?" line of reasoning that helps to explain mysteries right where we live.

Since we know the planets of our own solar system, we'll have to move Aeolus to another star. For the sake of convenience, so that we don't introduce too many startling contradictions to what we find familiar on Earth, we'll assume this star is about the same size and with the same characteristics as our sun, and that Aeolus orbits its parent star at the same distance—an average of 93 million miles—our Earth revolves about the sun.

That's where the similarities fade rapidly. Remember, Aeolus is a world of scientific imagination, a planet that never runs out of fascinating and bizarre wonders.

Our first view of Aeolus is from a distance of 239,000 miles—the same distance we find between the Earth and our moon. But

A WONDERFUL, IMPOSSIBLE WORLD

there's a difference. Orbiting about Aeolus is not only a moon but also two clusters of rocky debris. Each cluster is uneven in size, and they're believed to be a loose aggregate of meteoroid materials. One moves just before, the other just behind, the large lunar satellite.

Before we move on to Aeolus itself it might be worth taking another look at the moon, which in itself presents an extraordinary body. It should be what we call a dead world, for no life has ever moved across its surface. But there are many interpretations of what we mean by "dead" in referring to a large body in space. First of all, its mass is distributed unevenly. Far beneath the powdery surface there are mysterious concentrations of dense material that were never suspected before the first spacecraft went into orbit about the moon. Things happened that were "impossible." The spaceships orbiting the moon would, without warning or explanation, "dip" in their orbits, then return to their previous height above the surface so far below.

We use the word "impossible" because such an event defies everything we know about celestial mechanics—the rules that determine what happens in space. That's one of the great advantages of creating our world of Aeolus and its moon. By setting up "impossible" situations we're forced to search for the answers to questions we might otherwise never ask ourselves.

Is such a situation really so impossible—these strange dips and bobbing motions while in orbit about the moon? Well, if we can predicate such movement, we can also predicate the answer. What if—and we're making a case for theory here—what if there *were* huge concentrations of dense material far below the surface of the moon? Is such a thing possible?

There's one way to approach the problem. In the world of spacecraft leaving our Earth and orbiting our moon, the unusual dips and bobs in lunar orbit *have happened*. There was nothing to explain how and why such an event could take place. But by creating Aeolus and its moon, we have a so-called space laboratory in which to advance theories and solutions we might not care to try in the case of Earth and our moon.

Aeolus's moon has another unusual characteristic. When struck a great blow, such as would be created by a meteoroid rushing against its surface from space, the moon acts like a tremendous gong. It's a sort of incredibly large planetary bell that sends huge clanging echoes from the surface far into the moon and then bounces these echoes back and forth for many hundreds of miles. The meteoroid smashing into the moon acts as the clapper for the bell, and the moon itself is the bell that rings and peals its strange echoes and vibrations for many hours after being struck.

Time to leave the moon and turn inward toward Aeolus itself . . .

One of the lessons we learned within hours of sending our first satellite into space in 1958 was that we were incredibly ignorant of the conditions that exist in space. Not the distant stars, not the far-flung galaxies, but our frontyard space—the very space in which our world moved, in which it had always moved, and through which rushed terrible forces and radiations that affected everything we ever did or now do on this Earth.

It was a shock to discover that "empty space" was turbulent and seething with all manner of radiation. To understand what goes on about our own world, scientists have been using their "created planet" of Aeolus to establish clearer pictures of what goes on in space that we can't detect in any way with human senses. As we move toward Aeolus we study the blue-and-white planet through special screens that make visible to us the energies that fill the void between worlds.

It is a sight stunning, shocking, unbelievable—a look at forces beyond our wildest imagination. Just as we cannot see the waves of radio and television and radar but can create model images of what they are and what they do, so we are creating a model image through the special screen to bring into the spectrum of visible

A WONDERFUL, IMPOSSIBLE WORLD

light the vast radiation storms raging about Aeolus. Thus our mythical planet becomes a living model of Earth.

In one moment, looking at Aeolus through human sight, we are thrust into a fantastic vortex of energy that sweeps for hundreds of millions of miles in all directions. Pouring outward from its distant star is a howling maelstrom of energetic particles, a continuous blast of protons and electrons spewed forth in a raging stream that makes a tiny whisper of the greatest hydrogen bomb ever exploded. For this is energy on a scale so monumental that we can only look and gasp, and being able to see what happens to this energy as it cascades toward Aeolus teaches us a grim lesson—that the planet itself forms a magnificent barrier against energies so intense that it seems impossible for a single creature ever to have struggled to life on the surface of that world.

Our scientists granted Aeolus a miraculous feature—it is the densest of all the worlds in its system. It also rotates quite rapidly, with a speed along its equator of about a thousand miles an hour. Thus, through this density and this rotational speed, Aeolus becomes an extraordinary magnet, the forces radiating from the whirling planet so great that they create their own natural space barrier against the floods of seething radiation.

We look again through our special screen. Great sheets of blazing energy stream toward Aeolus but fail to reach the planet. Fifty thousand miles ahead of the shimmering globe, in the direction of the sun, there is a great, glowing band of blue-white light, so intense it seems almost to be the curving edge of an incredible mirror glowing along its surfaces and from within its own substance. This is a shock wave—literally—glowing brightly from a clash of energies that, were it to take place on the surface of Aeolus, would destroy every living creature within its reach. The shock wave is a line of battle between the outreaching magnetic field of Aeolus and the tremendous, continuous tidal wave of the stellar wind from Aeolus's sun. Here is the line of demarcation, where the clash is joined and where the vast energy outpouring from the sun is hurled aside.

PLANETFALL

Ten thousand miles closer to the surface of Aeolus we see another line of demarcation, what we call the magnetopause. Its edges glow a dazzling yellow-gold, a celestial fire flaming along the surface of a darker burning royal purple within. The magnetopause, forty thousand miles out from the surface of Aeolus, is the closest the intense magnetic field of the planet permits the radiations from the sun to reach the planet. On the inward side of the magnetopause, toward the world itself, is the magnetosphere, the "solid boundary" of magnetic force enveloping Aeolus like a shield of massive armor. The colors ripple and shimmer as the intensity of the outpouring radiation from the sun changes. At times, when the sun erupts with violent storms from its interior and along its surface, space is filled with an incredible interplay of blazing energies, all of them swirling about Aeolus and showing their enormous power in the form of a vast bowl of coruscating, brilliant color within which the planet is no more than a small, dense, spinning marble.

There are other bands of energy surrounding Aeolus, the streaming particles from the sun trapped in shifting bands of magnetic fields, but they are kept so high from the globe that they present no danger to the life existing on its surface and within its oceans. But we are not yet through, for something happens to the great rivers of energy flowing against and around the magnetic defenses of Aeolus. In all directions around the planet, moving backward along the boundaries created by the magnetopause, we see a gossamer fire, like a thinly glowing teardrop that extends for a hundred thousand miles above and below Aeolus, then spins and whirls back to form a magnetic tail that fans out, becoming thinner with distance, for more than a million miles behind the planet. It is a plume of utter beauty, a softly glowing veil through which the stars can be seen.

We switch off the special screen, and as of that instant it all vanishes, for now we are seeing, in the visible light, part of the spectrum. It is as if we have suddenly been struck blind to the most beautiful of all sights in space, and it is a wonder that not

A WONDERFUL, IMPOSSIBLE WORLD

until we went into space did we have any conception of these stupendous forces. And yet, we have been witness to only a portion of the massive energy flow hurled against the planet before us.

The star about which Aeolus orbits is hardly placid, for, like the sun about which our Earth orbits, it is a thermonuclear explosion going on steadily, with so much violence and of such magnitude that every second—*every second*—it rends and destroys four million tons of its solar substance, transforming that substance into violent energy. The slower-moving elements of this ceaseless holocaust are deflected, controlled, hurled away by the glowing shock wave and magnetopause of Aeolus. But what of the other radiations, many of which stream through space with the speed of light itself?

The magnetic shield created about the spinning globe of Aeolus is really only the first line of defense, for beneath the invisible belts of magnetic force is a thick atmosphere. It reaches to a hundred miles above the planet, although the densest part extends only a few miles above the surface. If we consider the diameter of Aeolus, some eight thousand miles, against the thickness of the atmosphere, it becomes obvious this layer of air is really on the order of onionskin protection for the world it surrounds. Yet it is tremendously effective in its role as a barrier, for its molecules, in effect, form a fluid medium through which solar energy must penetrate. And penetrate it does, for these radiations are what we consider to have extremely energetic and short wavelengths.

And the magnetic shield of Aeolus is helpless against this form of energy. The shock wave and magnetopause control the solar wind of photons and electrons because these particles are electrically charged and they obey the peculiar laws of magnetic forces.

Not so the ultraviolet and infrared and X-rays, and it is up to the comparatively thick (to us) barrier of Aeolus's atmosphere to dilute these forces. If they did not, searing energies would lash the surface of the planet. There is a constant "war" going on in the atmosphere of Aeolus. The powerful energy from the distant star smashes into the molecules of air, creating a constant floating zone high above the world of glowing charged particles. Through millions of years this floating zone has thickened and increased its density and provided the life forms of Aeolus with still more protection against the harmful elements of the mighty stellar furnace.

Still, some of the dangerous energies rip their way lower, for they are not affected either by the magnetic shield or the thick atmospheric barrier. Among what might be called dangerous would be the full strength of ultraviolet rays. Ultraviolet radiation of this sort, before it is filtered and reduced drastically in its strength, could so viciously flay exposed human skin that within seconds an unprotected person would be subjected to— many times over—a lethal sunburn. The scientists who created their model planet of Aeolus needed something that would stop ultraviolet radiation of such deadly magnitude. They reached into their hat of wonders and created an element they called ozone, and then spread it around the planet at an altitude of twenty-five to thirty-five miles. Convenient: their ozone layer stopped most of the ultraviolet radiation pouring into the atmosphere of Aeolus and reduced what was left to beneficial rays.

There are other wonders that are instantly made visible by more of the special screens through which we can view Aeolus. Looking at the planet from 200,000 miles out in space, through a modified ultraviolet filter, enables us to see another visual miracle—a geocorona made up of glowing hydrogen gas that reaches out to more than 50,000 miles in every direction about the globe. It is as if Aeolus itself were a completely dark sphere rolling silently through a fog that glowed and shimmered through all its substance.

We remove the filters and continue our approach to the planet growing larger and larger before us. Time for one more special screen for another magic look. This time the screen lets us see

A WONDERFUL, IMPOSSIBLE WORLD

the planet without the glow of hydrogen gas that makes up the geocorona. Now we see only the radiating energy from what the screen identifies as atomic oxygen and molecular nitrogen. Instantly Aeolus is transformed into a magician's crystal ball. One-half of the planet, that side facing the distant star, is a blinding white-gold light. The other half is in darkness, but only for minutes. As our eyes become accustomed to seeing through the special screen, fingers of gleaming golden light stretch from the lighted side to the darkened portion of the planet. The fingers begin to come apart at their edges and form swirling tendrils of gold and orange and red that whisper across the dark side of the world, curving around the edge. Now we see an interplay of radiation along just this part of the spectrum, and it is a sighing and gentle touch of color that is utterly beautiful.

The special screen is switched off. We are back to the visible light we have always known, and in some ways it is sad and disappointing to leave the magic we never knew existed. But there are other sights just as majestic, and they can be shared by all of us, and even captured on film. If nothing else, the minds that created the "experimental world" of Aeolus gave us the most beautiful of all the globes spinning through space.

The atmosphere of a planet is both fascinating and beautiful. It absorbs incoming visible light and then, through the mechanisms of that atmosphere—swirling gases, dust, water vapor, carbon dioxide, argon and other gases—diffuses and refracts light so that it assumes textures we might otherwise never know. Aeolus is a world to which a fascinating ocean of air has been applied. It is filled throughout with a great variety of substances in order that the scientists using the planet as their "think tank" might have great freedom of expression.

The clouds of Aeolus, for example, reach from the surface of the world to as high as sixty miles, but they are not the clouds we are familiar with here on Earth. The highest clouds of Aeolus are called noctilucent clouds, wispy wraiths of dust with frozen

PLANETFALL

water crystals whirled up from the lower reaches of the atmosphere. Here they glow softly, seemingly lit from within like a gentle flame brought to life by the distant star. But their gossamer hues are rarely visible, for not only do they ghost about the planet at such great heights, they move with almost unbelievable speed —hurtling along at 400 miles an hour.

Much closer to the planet's surface—but still far higher than the clouds we see as we stand on the Earth—are perhaps the most beautiful clouds we could imagine. These are the magiclike mother-of-pearl formations, ghostly nacreous clouds of ice crystals, gliding swiftly and silently through air that is less than one percent of the pressure found at the level of the sea twenty miles below.

The other clouds of Aeolus are much like those beneath which we live our lives: cumulus, nimbus, stratus. Thunderstorms on this world, however, reach to enormous heights, as great as 75,000 feet above the land, and they boil with energies that could easily destroy the strongest airplane ever built. Seen from space, the thunderstorms of Aeolus (especially at night) are like huge electric-light bulbs flickering almost constantly with threadlike tongues of fierce lightning. And to be sure that the atmosphere of the planet is always electrically charged, the scientists established that there should be at least fifty thousand thunderstorms going on at any one time around the world.

Just to add spice to the situation, and to be certain (for some manner of scientific experiments) that there will always be enormous levels of electrical energy, the scientists created lightning of special violence for Aeolus. On that world the lightning bolts at times stretch for many miles through the air: they are thick, often braided in their form; they move with the astonishing speed of 26,000 miles a second (which would take a single lightning bolt around Earth, along the equator, in less than *one second*), and they burn with a temperature of 45,000 degrees.

Now, when we consider that the surface of the sun is only

10,000 degrees, we begin to imagine the staggering electrical storms that rage across Aeolus!

Any ocean of air, if it is to be host to advanced forms of life, must itself be complex and intricate in its materials and in its effect upon its world. This situation applies, of course, to Aeolus, where scientists exceeded themselves in "creating" on one world a variety of life forms that really should belong to a hundred different planets. If they extended themselves with the magnificent electrical and magnetic properties of Aeolus, then they went beyond normal life forms to the monstrous, with which they populated their startling world.

Again—and every so often we must provide ourselves with this reminder—the "creation" of Aeolus was made so that we might better understand the mechanisms of developing life, the extremes to which that life will reach, and what might kill off whole species without warning.

But to get that close a look at the fauna of Aeolus it is time to land—to make planetfall.

CHAPTER 3

Planetfall on Aeolus

It's safe enough to be on the surface of Aeolus, *if* you choose carefully the time and the place. On this world of great extremes we have to exercise the greatest caution. Landing in an unknown area could be fatal, if for no other reason than that most of the planet is covered with oceans. The reasoning is sound—the oceans contain the chemicals and the nutrients and the protection necessary for life forms to begin, to evolve into higher forms of life. On Aeolus, nearly eighty percent of the world is water. What might well be called a liquid planet.

That leaves just a bit more than twenty percent of the total surface of the world for land, but the figure is misleading. If you want a "think tank" world on which you can run your experiments, you've got to fill the available land surface with a group of extremes of conditions and climates, even if they clash so severely that it's hard to believe such a place could exist.

For example, the temperature extremes seem ridiculous. In the hottest portions of Aeolus, temperatures recorded in direct sunlight reach to as high as 155 degrees—or to what we would record as 155 degrees on the Fahrenheit scale.

The other end of that fearsome thermometer would be at just about 100 degrees below zero, again on the Fahrenheit scale. But that's only a part of the story, for temperature has to be judged accurately on the basis of what we call the chill factor. When you get down to below freezing, or below zero degrees F., you have to consider the wind. Using a rough rule of thumb, the *effective* temperature drops one degree with every mile an hour of wind. How high are the winds on the *surface* of Aeolus? In the cold climates, where the surface is hidden beneath snow and ice, winds of nearly 200 miles an hour. If it's 40 degrees below zero and the wind is doing, let's say, 170 miles an hour, then your effective temperature is 210 degrees below zero.

It's difficult to look upon Aeolus, then, as being hospitable, because we're now considering the range of temperature to be 365 degrees! From 155 degrees above to 210 degrees below zero.... Can you imagine having to live on an Earth like *that*?

From the lowest point of land to the highest on Aeolus there is a distance of some thirteen *miles*. That's a stretch of more than 65,000 feet.

Vast areas of the land mass—despite the shortage in real estate—are not only inhospitable but menacing to all forms of life. There are deserts that for hundreds of miles are nothing more than sandy, trackless wastes where no rain ever falls and where nothing lives or *can* live. Other deserts are bare rock. Others are vast fields of ice and snow, perpetually frozen. There are huge mountain areas, so forbidding that they should be shunned by all life forms. Great swamps and deltas occupy much of the land surface. In short, the land makes up only a little more than twenty percent of the entire planet, and of that area, far more than half is inhospitable and even threatening to higher forms of life.

We can see, then, that the "think tank" world of Aeolus restricts the development of advanced life forms to isolated patches and strips of land. That may not appear to be anything near generous even to a scientific experiment, but it does cut down on the complications of keeping records of different species.

In fact, Aeolus is really two completely separate and distinct worlds on a single planet. One is the world above the surface of the oceans (to which we'll return shortly), the other is the hydroworld made up of the vast liquid content of the globe. (Maybe the scientists should have named their special project Aqua rather than Aeolus.)

PLANETFALL

You've got to have some system of measurement and some selected figures to use within your yardstick. The figures for Aeolus came to a total of 140 million square miles of oceans. An impressive figure, but strictly for the surface of all that water. Working out the shape of the bucket to contain all those oceans brought forth a figure of 329 million cubic miles of liquid for the planet.

Because oceans are great breeders of life, and they act as engines for the mechanism of weather, and they reflect the characteristics of a planet's crust, Aeolus needed more specifics than those already listed. The scientists said that if the global oceans were an average of 13,000 feet in depth, and that four-fifths of all the oceans were deeper than 9,000 feet, then all the oceans and the seas together would contain the staggering mass of 1.6 billion billion tons of water. This also led to some interesting juggling of figures. On such a world, if you rounded off the surface until it was completely smooth, the water covering *everything* would reach a depth of nearly two miles.

Enough of weighty statistics about the water world. Anything that might be said of the physical characteristics of Aeolus must pale in comparison with the astonishing variety of life forms on the globe. We said earlier that Aeolus would be a biologist's paradise.

Imagine so great a variety of flying life forms that the largest winged creature is more than a billion times heavier than the smallest! On Aeolus, scientists can do just about anything, and they worked out that such a thing was possible. To those of us familiar with life on Earth, it sounds completely ridiculous.

Other winged creatures fly to heights of 25,000 feet above sea level, which would put that particular species of bird at an altitude where a man would collapse within ten minutes or so from lack of oxygen. Still another Aeolusian species is capable of tremendous diving speeds, being able to reach a velocity of about 360 miles an hour in a screaming plunge toward the ground.

Within the oceans of Aeolus are even stranger creatures, where

PLANETFALL ON AEOLUS

the largest is several billion times heavier than the smallest. The largest animal on this world is about 120 feet long and weighs nearly 400,000 pounds. Again, keep in mind that this is an intellectual and scientific exercise, and while such creatures may not be probable, we can argue that they are at least possible.

This 200-ton animal is neatly divided into several separate sections in order to function much like a biological machine. One-third of its body is actually a massive engine of muscle power—and you need a lot of power to ram 200 tons of body through the seas. The animal is capable of maintaining a speed of nearly twenty-five miles an hour through the ocean, which means, converting muscle power to mechanical figures, that it must churn out the equal of some 900 horsepower as it pounds through the waves.

Within the ocean depths are even more startling creatures. There are deep-sea eels that reach a length of more than ninety feet. There are shrimp that bob along unconcernedly at a depth of seven *miles,* where the pressure on the body of such creatures is measured at *seven tons to the square inch.*

When you have a whole world to manipulate you can come up with creatures so astonishing that every one of them seems a test of wild imagination. There are sea creatures listed for the oceans of Aeolus that move freely thousands of feet beneath the ocean surface, where they communicate with one another by intense bursts of light, of many different colors, flashing their lights on and off in rapid, intelligent sequences. A few others include fish with long silken tendrils extending well before them, with dazzling lights at the end of the tendrils. The Earth has its familiar squid that releases a dark inklike substance, behind which it hides from danger. On Aeolus there are squid that eject huge clouds of blazing, burning particles that turn the darkness of the deeps into wildly sparkling wonderlands. All colors are recorded: blue, blue-green, green, white, orange, red, purple-orange, pale green, yellow, and variations of these hues.

Because the "making of Aeolus" created so many areas of

harsh and what would normally be lethal environments, the scientists were forced to produce some far-out solutions to the problems that otherwise might have wiped out their carefully designed forms of life. Take different types of fish living in polar waters so cold that the fish *must* die. But they don't. The answer? Simple, explain the scientists. We use Prestone to keep a car engine from freezing; they "invented" a natural form of antifreeze for their fish, and called it glycoprotein. When the water temperature gets down to body-freezing levels, the fish merely produce more glycoprotein and alter their biological systems. We could say it's mighty convenient.

Here's a partial list of some of the more "imaginative" life forms created for Aeolus:

Bacterialike creatures that live quite well in water heated to temperatures of 170 degrees F. above zero.

Trees that survive handily in temperatures down to 90 degrees F. *below* zero. There are also trees that live for thousands of years and other forms of vegetation that flourish where rain never falls and where the humidity is about as close to zero as one could imagine.

A variety of mosses, algae and lichens that can survive years of exposure to temperatures of 400 degrees F. below zero (no doubt to be available for new life forms in the case of a dandy planetwide disaster).

Other plants that can breathe oxygen when the outside pressure is equal to the pressure we would find above our Earth at a height of 100,000 feet—which is just about empty space.

How about a specially designed plant that can live *without any air*! The scientists put their heads together and came up with what they call anaerobic respiratory systems. These plants need sun, and that's all. By receiving sunlight, they carry on their process of photosynthesis and produce their own oxygen within their systems. Neat.

They even designed plants that can live within a vacuum by growing extrastrength outer surfaces that make a sealed pressure system of themselves. It's handy, all right. The plant roots in almost anything, grows in a vacuum, and produces what it needs within its sealed system by absorbing sunlight.

A form of reptile that resembles a turtle but has a built-in adaptability to almost any external conditions. Such a creature could be taken from normal sea-level temperature and pressure and placed on top a mountain 55,000 feet high. Even if the blood volume fell to the point where it would be impossible to detect any signs of blood moving within the reptile, it would survive.

Radiation is always considered both a boon and a danger to life forms. The scientists created a lichen (they said it would resemble, in appearance, the reindeer moss we know on Earth) that could withstand as much as four thousand times the ultraviolet radiation that reaches the surface of Earth. By every rule of science with which we're familiar, no living form can survive under ultraviolet radiation of only a fraction of this intensity.

What about the more penetrating radiations like gamma rays? The special moss of Aeolus was "designed" to withstand a dose of gamma radiation a thousand times deadlier than the radiation that could kill a man—and the moss would keep right on growing.

They also "produced" a small many-legged creature—the closest description would be the cockroach we find on Earth—that could endure a *million times* the atomic radiation that would kill a man, and it would still be unaffected. Maybe they want to be sure that if there's an atomic war on Aeolus there will still be many living creatures around, no matter how intense the radiation levels.

Finally, the scientists shaping their planet of Aeolus came up with two special characteristics of their private world that should make us regard the whole affair as an exercise in fantasy rather than imaginative science.

They "created" two "miracle substances" that would enable them to shape the planet in any way they wanted. One, an essential element of the atmosphere, would be a gas that would be clear, odorless, nontoxic and that would conform to the physical laws of a perfect gas. It would always retain its chemical composition and would remain fluid at all temperatures. The second miracle substance would be an incompressible, stable liquid that could easily and quickly be converted either to a solid or a gas, and could just as quickly be reconverted to its original state. And it would also have to be nontoxic in all its states, and easily stored.

One last item was prepared for Aeolus, and, obviously, it was another "miracle-maker" that, by its existence in the chain of life of the planet, would solve any nasty problems that might arise. This would be an atom that could mix easily with other atoms and would also be an "incredibly efficient energy container." It would exist as an atom in a state by itself, or could be coupled in molecules, or even bonded triply in more elaborate molecules. But its biggest value would be the extraordinary manner in which it released energy.

We're familiar (back on Earth) with a process called fermentation—a chemical process that is able to sustain life. Fermentation, on the average, releases fifteen calories of energy for every gram molecular weight of its substance.

The "miracle atom" would have to do a lot better than *that*. To get away with all the incredible conditions of life they postulated for Aeolus, the scientists stated that the miracle atom (on which all animal life would depend on that world) would have to release 674 calories of energy for every gram molecular weight of its substance. Now, compare *that* to fermentation!

There's also——

Well, enough of Aeolus.

CHAPTER 4

Third Planet from the Sun

Have you caught on yet?

Did some of the miracles of the planet Aeolus seem to be just a bit familiar to you? Well, there would be some familiarity, of course, because in the scientific conception of such a world we would use the Earth as a starting point for the model. This imaginary planet was named for Aeolus—in Greek mythology, king of the winds on Earth. But the planet Aeolus—under a different name—*is real.*

It exists. It's been around for somewhere between five and six billion years. Life has stirred within its oceans and on its surface for perhaps three billion of those years.

It's the third planet from the sun. Between Venus and Mars.

The planet Earth.

Everything you just read about this fascinating world, no matter how incredible or startling, is true. Every word of it. Of all the planets we might imagine, of all the varieties of "impossible" life forms we might bring to mind, of all the magnificent sweeping energies we could try to create—we can do no more than to describe only a few of the truly marvelous wonders of our home planet.

Perhaps this is the best way to gain a better appreciation of just how startling and incredible is the planetary home of man. Sometimes we've got to look elsewhere—or at least seem to do so—to "rediscover" the wonders all about us. Earth, in its descriptions as Aeolus, seems not only far-fetched but impossible.

PLANETFALL

Let's look back at some of those "impossible" facts and situations.

Early in chapter 2 we stated that Aeolus not only had a large moon but was also orbited by "two clusters of rocky debris." Well, every schoolboy *knows* that the Earth doesn't have any such thing, so the two worlds—Aeolus and Earth—can't possibly be the same. The truth is that the Earth *does* have two cloudlike satellites; they can be observed with the naked eye under only the most favorable seeing conditions, but they can be detected by telescope and they can be photographed. Each satellite is actually a cluster, or swarm, of tiny meteoric particles. They orbit the Earth at just about the same distance as our moon, approximately 240,000 miles, and trail the moon by some sixty degrees. And they *were* photographed, the first times on March 6 and April 6, 1961, by Dr. K. Kordylewski, of Krakow Observatory, Poland.

They're real, all right, even if hardly anybody has ever heard about them.

Then we discussed what was supposedly the moon of Aeolus. Keep in mind that from this point on everything we say about Aeolus/Earth is true. We referred to "mysterious concentrations of dense material" beneath the lunar surface that caused spacecraft orbiting the moon to dip suddenly in their orbits and then return to their predicted paths. These, too, are as real as everything else. They're called mascons (for mass concentrations of matter), and they have now been mapped on the side of the moon facing Earth, as well as the far side that is never seen by a viewer standing on our world.

Earth's moon was described as acting like an "incredibly large planetary bell" when struck by a huge meteoric particle. The statement is true when the moon is struck by large rocket stages, such as the Saturn S-IVB boosters that sent our Apollos to the moon, and were then directed by radio control to impact on the lunar surface. The "huge clanging echoes" were one of the most unexpected effects found in the extensive research carried on about the moon.

THIRD PLANET FROM THE SUN

Everything you read about the enormous sheets of energy lashing the Earth (Aeolus) is true except the "special screens," and I used these only to emphasize the magnitude of the radiations that sweep through space. Everything else—the enormous magnetic field of Earth, the great shock wave fifty thousand miles ahead of our planet—all of it, it's true.

But what about those strange, truly incredible clouds that were described? The noctilucent and ghostly nacreous clouds that race high over Earth. Can they be real? They are, and they have been tracked by radar and photographed by special cameras. Sometimes, when the sun angle is "just so" at dawn or dusk, we can see them with the naked eye. It's just that most people who catch sight of these filmy wraiths rushing so far above our heads have no idea what they are. But they're real.

We described thunderstorms that reach to heights of 75,000 feet, and are so mighty they can be seen from space, "huge electric-light bulbs flickering almost constantly with threadlike tongues of fierce lightning." But we know that's impossible, that the greatest thunderstorms never climb higher than about 45,000 feet.

The truth is that we believed this to be so, that the upper limit of such storms was about nine miles above the Earth's surface. When men stormed into space to orbit our planet we learned just how wrong we were, for they were able to look down on the mighty storm giants arrayed against distant horizons, and with special instruments measure their towering height. The description of the thunderstorms as seen from space? At least eight astronauts have described these to me personally, and even the first American to orbit the Earth, John Glenn, reported his awe at the fabulous sight.

What about the bolts of lightning that move so fast that a single bolt could girdle the entire planet along the equator in less than one second? And the temperature of 45,000 degrees? It's true, and the next time you see a lightning bolt, keep those figures in mind. We've been treated to these dazzling displays of energy all our lives without truly realizing what we were seeing.

In chapter 3 we moved closer to the surface of Aeolus/Earth, and here I adjusted the figures slightly to keep up the deception. Aeolus is described as nearly eighty percent water. If we're discussing Earth we need only to change that figure from eighty to seventy percent and we're dead on target.

The temperature extremes of Earth? Just the way you read them. We referred to winds of "nearly 200 miles an hour." On *Earth*? Absolutely. Years ago, wind forces of 188 miles an hour were measured right in New Hampshire, and the shrieking winds of typhoons in the Pacific Ocean have exceeded 200 miles an hour. If anything, I confess to being conservative with these figures.

What about the height from the lowest land to the highest, where I used a figure of thirteen miles? If you measure the land *beneath the ocean,* then you start about seven miles down, in the Challenger Deep near the Marianas Islands in the Pacific and you keep going up until you reach the top of Mt. Everest, just about six miles above sea level. The ocean simply covers this awesome spread between the two extremes.

All the figures you read on the hydroworld of Earth— 140,000,000 square miles of oceans, 329 million cubic miles of liquid, and the rest—represent accurately the statistics of our planet's vast and deep seas.

We went on to life forms where we read about the largest winged creature being a billion times heavier than the smallest thing that flies. And you have the albatross and the gnat.

Birds *do* fly to heights of 25,000 feet, and most likely they soar even higher. What about a bird that can dive at speeds of 360 miles an hour? They exist in the hawk family.

Even these extremes seem acceptable when we came upon the largest animal of Aeolus/Earth as being 120 feet long and weighing 400,000 pounds. And being able to generate 900 horsepower to ram its bulk through the waves. We're talking about the largest creature on Earth today, or that ever existed on our planet—the great blue whale.

But can there really be sea creatures, such as the shrimp that was mentioned, that can exist so far down in the ocean that the pressure on their bodies is seven tons to the square *inch*? Again I moved the figures around. The facts are that red shrimp were discovered at the bottom of the Challenger Deep, by the bathyscaph *Trieste,* at a depth of 35,800 feet, where the pressure is *eight tons* to the square inch. Not only shrimp; scientists aboard *Trieste* also watched a thirteen-inch flat fish, with both eyes on one side of its head, scurrying along the bottom.

The deep-sea eels of ninety feet? The results of ocean research by scientific vessels of the Danish Navy.

We move from the bizarre to the impossible when we describe deep-sea creatures that "communicate with one another by intense bursts of light, of many different colors, flashing their lights on and off in rapid, intelligent sequences." That description applies directly to what scientists in bathyscaphs have observed with their own eyes far within the ocean. And the squid that "eject huge clouds of blazing, burning particles . . ." They're as real as the shrimp moving along under a pressure of 16,000 pounds to the square inch, and the great blue whale that weighs 400,000 pounds.

When I described fish in frigid seas that manufacture their own form of Prestone—glycoprotein—I was describing exactly what scientists have reported from the Antarctic.

And all the rest—trees that survive in temperatures of 90 degrees below zero, other mosslike forms that withstand 400 degrees above zero, plants that seal themselves off and manufacture their own oxygen within a sealed, pressurized environment, a turtle that survives at 55,000 feet, lichen (reindeer moss) that can withstand savage ultraviolet and gamma radiation, a cockroach that can endure a million times the lethal radiation dose for a man . . .

They're all real.

But what about the "miracle substances" such as the gas that would be "clear, odorless, nontoxic and that would conform to the physical laws of a perfect gas," that "would always retain its chemical composition and remain fluid at all temperatures . . ."

That is a concise description of *air*. The very air you breathe.

That second "miracle substance," described as "an incompressible, stable liquid that could easily and quickly be converted to either a solid or a gas and just as quickly be reconverted to its original state," and was "nontoxic in all its states, and easily stored . . ."

Know what *that* one is?

Water.

Last, but obviously anything other than least, was the "miracle atom" that was "an incredibly efficient energy container . . . upon which all human life would depend."

The miracle atom that, like water and radiation and a dazzling host of other factors, is responsible for the existence of all of us. Oh, it's real, all right, just the way it was described.

It's known as the oxygen atom.

Chapter 5

The Splendid Destiny

The machine rushed toward Earth with an incredible speed: three hundred miles every minute. High above the surface of the world the lone man within the gleaming sphere looked down on a vast globe mantled in darkness. He turned to the east; there the dim, curving horizon showed a small deep-red crescent. Swiftly, as he rushed toward that dawn at five miles a second, he watched the dark-red color yield to rich red-orange tints, then increase its brightness to dazzling yellow-orange. The horizon emerged from blackness to a widening strip of magnificent royal blue, then yielded to a sliver of intense white cresting the edge of the world. The colors began to blend, the layers easing smoothly into a single expanding pattern of delicate hues to witness the swiftest dawn ever seen by man. Still, the intensity of light grew, and then, with shocking impact, the sun exploded silently into view from the other side of the planet.

The race of man had passed through its first dawn in space, and knowledge of this quiet, distant event would change forever the role of mankind. The sphere rushed over continents and oceans and then it was time to begin the first planetfall ever. Flame stabbed vacuum; billions of sparkling motes flew wildly away from the gleaming sphere. It curved downward more steeply and, like a blazing meteor, hurled itself into the thickening soup of the atmosphere. In a long shriek of fire the sphere reduced its enormous velocity. Minutes later it eased through clouds, drifting gently beneath a great billowing parachute. Finally, it touched rich farmland.

PLANETFALL

The first planetfall.

In years past, explorers who ventured forth from the world familiar to men journeyed across the great seas of the planet. New worlds then were greeted with the jubilant cry of "Land ho!" And when men stepped forth from their ships to plant their flags on soil they had never before seen, and the existence of which most men doubted, they altered forever the destiny of their nations.

When Cosmonaut Yuri Gagarin returned to Earth after the first orbital flight of man on April 12, 1961, he began a chain of events that would, both in subtle and in drastic ways, change forever the destiny of the human race.

For the first time in the six billion years of Earth's existence, a creature had appeared and torn himself loose from the planet itself. Until that moment, evolution gripped in an iron fist the destiny of all things that walked, swam or flew on Earth. Every creature from time immemorial was an absolute slave to the crushing gravity of the planet. Now, all that was over. Now, a new wave had begun, and it would continue outward, until it touched other worlds.

The upward climb began perhaps two million years ago. While we have journeyed the cindered dust of the moon, men of science still debate vigorously with one another in their arguments as to the origin and age of man. If, however, we push to one side the many arguments, we may say with strong reason that man first appeared in a recognizable form two million years ago. Five hundred thousand years ago, modern man emerged from the tumultuous experimentation of nature. *Homo sapiens* in his most distant form, the far end of ancestry belonging to one and all of us, began to thrust himself upon the world.

On the scale of evolution of mighty creatures that once roamed this planet, man has been here for hardly more than the whisper of scant seconds. In the Mesozoic era of evolutionary life it was the reptile that dominated the globe. Huge and magnificent creatures that stormed the depths of the sea, shook the earth with

THE SPLENDID DESTINY

their ponderous movement, rushed through the skies. For 155 million years the reptilian world possessed all of this planet.

One hundred and eighty million years ago the first mammals moved furtively onto the scene of evolution. It took the mammals another 110 million years, for whatever reasons there were, to finally displace the reptiles as the leading form of life.

For the past 70 million years the mammals have been dominant. And not until very recently, on such enormous scales of time, did man appear as the species *Homo* until, a half million years in the past, intelligent man—*Homo sapiens*—began his domination.

How far back did modern man shoulder his way into the global scheme of things? It may be written down as only 25,000 generations. But those were the *first* men, and they lived as animals, changing slowly in their evolutionary path.

Think of this incredible mastery of a world, and then beyond, in this single fluttering heartbeat of evolution. Forty thousand years ago, fighting the overwhelming disaster of the last great ice age on this planet, Cro-Magnon appeared suddenly. The creature the anthropologists call man had been here for perhaps two million years, and he had fashioned crude weapons and tools, but now there was a newcomer who painted on the walls of his caves. A new creature who had the ability of thought that so towered over all other creatures that he was capable of reason and logic. He was a creature who overcame his fragile strength compared to the animals roaming his world, and, with the awareness of his powers, went on to assure his mastery of all beasts, no matter if they swam or walked or flew.

This creature, the new man, could do what no other living thing had ever done. He learned to organize the resources about him, and, in so doing, he brought forth a creature unlike any other, in yet an even more meaningful way.

He would become the master of his world, and, in the process, he would be able to dictate his future, rather than stumble blindly into an unending succession of murky tomorrows.

PLANETFALL

One hundred eighty million years ago the first mammals appeared on this planet.

Two million years ago the first man-apes emerged.

A half-million years ago a new creature evolved from the man-apes. The first men were here.

Forty thousand years ago, struggling to survive in the midst of the last great ice age on this world, Cro-Magnon came upon the scene.

Twelve thousand years ago man, as he lives today, began to farm, assemble his cities, forge his weapons.

Seventy years ago man took to the air in a machine that flew under his control.

Less than thirty years ago man tapped the secret of the atom and unleashed the energy of the great suns that fill the universe.

It has not been quite thirteen years since the first man returned from the space beyond Earth, and made the first planetfall.

Eight years and three months after that date, the first man descended a metal ladder and set his foot on another world.

Think of this incredible shrinking of time.

Two thousand generations separated Cro-Magnon cave artists from the first men to struggle into the air.

Everything else we know today has happened in just one generation more.

It took forty thousand years from Cro-Magnon to Kitty Hawk. Forty thousand years to reach a flight that lasted twelve seconds, that spanned a distance of 120 feet.

Sixty-six years later a great ship sailed away from the Earth and headed across the sea of vacuum for the moon, a quarter of a million miles away.

In 1903 the first Wright brothers airplane took 12 seconds to fly 120 feet.

In 1969 Apollo XI took 12 seconds to sail *84 miles.*

And to take three men far enough for two to make the first planetfall on a world other than Earth.

How far away is Mars? Venus? Jupiter?

THE SPLENDID DESTINY

On this scale of life, against the enormous surging power within the grasp of modern man, not very far at all.

Sometimes it is necessary to stand far back from where we normally study things in order to gain a clear look at what we're trying to see. There's an old expression that often you can't see the forest for the trees. The reasoning applies to us here on Earth, as it does as far into space as our vision and imagination can take us.

We know there are nine planets moving in their majestic orbits about the sun. We *know* this to be so because, through one means or another, we have seen or detected these globes and charted their ponderous moves along the celestial spider web of gravity and centrifugal force. But not too long ago we *knew* there were only six planets. It wasn't that far back in time when we *knew* that the planet Mars lacked any moons, or that Jupiter had only four moons (rather than the twelve we now know to circle that huge world), or even that space flight was impossible. And it hasn't been that long since we've had to change our textbooks to read that Saturn really has ten moons instead of nine.

Which brings us to this point: When we talk of Earth we've got to specify *which* Earth. The one we knew before the first satellite was rocketed into orbit on October 4, 1957, or the drastically changed world since then.

Let's hold it right there a moment. The world hasn't changed that much since 1957, of course, but we have learned so much, that the planet Earth in the age before we had satellites whirling about this globe really seems to be a world almost strange to us.

Which brings us to one of the most important issues of man and his future. Why bother to venture away from Earth?

To me the answer isn't found in the loaded clichés that we've heard for so many years. One of the pet phrases regarding this question is that we're "exploring space." We aren't doing anything of the sort, any more than Ericson or Columbus risked their lives to explore the *ocean*. The great seas of Earth were a means of getting from here to there. For whatever reason—to find a better trade route, to search for riches, or simply to answer an insatiable curiosity, or perhaps all of these reasons and a thousand more—men raised their sails and set their course for distant and uncharted lands. And every time they did so they widened the horizons of all mankind, they added to the treasure house of knowledge, they improved the ability of man to control the environment about him. Which is another way of saying we were adding to our ability to affect the future that would roll inexorably upon all men.

Can you imagine the world of just about seventeen years ago, when the first satellite had yet to leave its launching pad? In those days we were blissful in our ignorance of vast and terrible forces affecting us every day of our lives.

Just as the men of ages past were prisoners of their lives, so we were unknowing prisoners. There existed a fascinating and incredible flow of energies and forces about us. Ignorant of their existence, unable to "see" beyond our limited vision, we continued to stumble into the future.

The "good old days" you hear about so often are the result of stunted memory. There are moments of the past to cherish because of their charm, their rewards, their challenges. But to cling to the past because it seems superior to today's problems is to run from reality.

Some of Earth's problems today seem overwhelming to us. The global ecology has been battered. Pollution affects us all. Population growth keeps hundreds of millions of people in near starvation, and millions indeed do starve, despite enormous increases in food production in almost every land on this planet. We suffer one raging small war after another. It's a terrifying list.

But what about yesteryear? Without antibiotics or the miracle medicines or surgical procedures that keep so many people from being condemned either to continued agony or an early grave? Does anyone really prefer surgery without anesthetics, or the excruciating pain of the dentist's chair of fifty or a hundred years

THE SPLENDID DESTINY

ago compared to modern dental science? Heart surgery was an impossible dream. Organ transplant was fantasy.

Unless you were wealthy and had years of time to spare, travel on a regular basis was reserved for the very few. Modern technology brought us gracefully swept wings and mighty jet engines and opened the world to us all. Not so long ago it took longer to travel a few hundred miles than it does for us to journey in upholstered comfort, with far greater safety, from London to Tokyo. That's a world of fascination and it's here to be enjoyed.

We accept live television broadcasts from any part of this planet to another as simply a fact of life. But that's strictly a result of communications satellites hurled into high earth orbit, a dividend of the space age that only twenty years ago most people regarded as ridiculous, impossible, fantasy, or a collection of all these attitudes.

It is surprising how many people bemoan the outbreak of wars around this planet. No one in his right mind is going to advocate war as a desirable solution to the problems of man, but wars are not new. In the past four thousand years of recorded human history there have been perhaps 230 years without a major war of some kind. You get the impression after a while that—right or wrong—men accept war as a natural way of life. And it's easy to forget that the majority of these wars haven't been carried out by the major powers but represent highly charged emotions on the part of small nations, tribes, groups, whatever, all trying to stake their claim for a piece of the action. The solution to this problem hasn't changed from three thousand or three hundred or thirty years ago; it has nothing to do with technology, and it has everything to do with man. War isn't a creation of nature. People —not nature or science or engineering or technology—make war. Cursing man's science as the culprit is like baying at the moon. All you get for your pains is a sore throat.

But one of the dividends of learning to place our electronic wizardry in the space environment *has* done wonders to stave off the nuclear war rightly feared by all men of common sense. A

PLANETFALL

dazzling variety of military satellites, none of which carries so much as a crossbow as a weapon, today circles our planet at heights ranging from ninety to 60,000 miles above Earth. Their purpose? To deny any nation the incalculable advantage of surprise in a commitment to major war. How? By keeping the entire globe under constant surveillance with special instruments that can detect the explosion of a nuclear device or the launching of a great missile, at any time of day or night.

The fact that there exist today, in the hands of half a dozen great powers, something on the order of two to three hundred thousand atomic and hydrogen bombs, and that we're not engaged in smashing cities and slaughtering whole populations, provides some rather serious food for thought. Somewhere, somebody is doing something right. And if these wonders of science that gaze relentlessly down upon this planet from their enormous height have helped to make that possible, why, that alone is worth every dime we have ever spent or will spend on what we curiously call the "space program."

But all this, no matter how important, still sidesteps the main issue. Once we sailed the oceans of our world to gain not merely access to new lands but to open wide the horizons to a future far more wonderful than we might ever dream. There's always an attempt to judge in the economic sense the pursuit of knowledge. Many scientists decry this and insist that scientific research has never been amenable to rigorous cost accounting in advance. This also is true of exploration of *any* sort.

Both sides have their valid point of contention. Unbridled expenditures can wreck the economy of any society, and science then becomes a bitter taste in the mouth of one and all. But it's also true that a nation, a people, who refuse to commit some of their energy to the future somehow end up by being mired in the present.

And the present has a nasty way of swiftly becoming the past. Time, progress, growth—call it what you will—have no patience

THE SPLENDID DESTINY

for those who can't see farther than the end of their economic noses.

Experience is, or should be, a teacher worth remembering. And experience with growing capabilities in science and technology teaches us that the future is fairly bulging with a treasure of unpredictable benefits in economy, industry, science, medicine, communications and—well, make a list of just about *everything* important to the race of man and it's included.

The past dims quickly and there are times when we have to force ourselves to understand the crippling limitations man has long imposed on his own growth. One of the most dangerous of all possessions of man has been knowledge. Call it the truth about life, if you will, for it's really one and the same. Yet some of the greatest minds we have ever known were put to death because their possessors had learned, or discovered, and spoke the truth about their knowledge.

This is not always the way to win popularity contests. The upper hierarchy of any society usually is extremely fond of its advantageous position. Anything that rocks the boat, and especially truth about life, is to be destroyed at once. Sounds impossible, but——

In the late 1500's the brilliant Polish astronomer Copernicus allowed his findings to be published, in the form of evidence arising from exhaustive study, that what almost everybody in the world believed—or *knew*—about their planet was woefully in error. Man, Copernicus seemed to be saying, was marching through life in appalling ignorance of the most basic facts of existence. He wrote that no matter what people preferred to believe, the sun truly was a far greater body than was the Earth. He also said that it was the Earth, and all the other planets, that moved in their paths about the sun, not, as people insisted, that Earth was the center of all creation, and that everything moved about *us*. As for the stars, they were really great suns incomprehensible distances from our system of one sun and its planets.

Copernicus never did find out the reaction to the published

PLANETFALL

evidence of his astronomical studies. Apparently he was as good a student of the politics of the day as he was of science, for he had let it be known to a very few close friends that his statements were sure to lead either to torture or to death. His material was published as he was on his deathbed.

No one could doubt his concern for his own well-being had he published his work while still alive. Shortly after the book of Copernicus became available to other men of learning and science, the Italian philosopher Giordano Bruno, using Copernicus as his fount of knowledge, went on to say that Earth could hardly be so unique as to be the only world of its kind in all the universe, and it was probable there were many globes like our own, circling other suns, on which there must exist intelligent beings.

Well, that was *too* much. In the year 1600 Giordano Bruno, arrested eight years earlier, was dragged before the outraged leaders of the Holy Roman Inquisition, who demanded he retract his vile words. Bruno refused.

So, naturally, they burned him alive.

Which, undoubtedly, helped George Bernard Shaw to arrive at the conclusion that he did not know what the inhabitants of the other globes were doing but he was firmly convinced they used our Earth as a lunatic asylum.

Well, if we can get over the walls of our asylum, let's wrap up our thoughts on Earth before we embark on a journey through the solar system. Keep in mind that any trip through the enchanted loom of centrifugal and gravitational forces (which keep the planets neatly in their orbits and prevent rather a mess) is just as much of an exciting adventure as it is a scientific expedition. The two quite often go hand in hand, which is icing on the cake.

In earlier chapters we learned that Earth is not isolated in its place in our solar system, or as part of the Milky Way Galaxy, but plunges through a vast and seething storm of terrible radiations only recently "discovered" by us. Modern history is filled

THE SPLENDID DESTINY

with repeated discoveries of forces of which we were ignorant. We didn't need to leave Earth to discover *that,* of course. Radio waves have been with us as long as the stars have blazed, but we didn't know that until we found the means to detect such radiation and then built our own devices that transmitted carefully controlled forms of energy for radio, radar, television and other useful purposes. But radio waves, in all their forms, still are only one tiny splinter of the forces that lash through space.

The *real* mystery is gravity. No creature has ever evolved except under the total subjugation of gravity. Yet it baffles us. At one time it appears to be both the most powerful force in the universe and also one of the weakest. Gravity is the loom from which universal order is spun, yet a child's magnet snatching a pin from a table has just overcome the gravitational pull of this entire planet. Gravity pervades every iota of our lives, and yet no scientist can tell you what it *is*. Science can measure its force, predict what it will do, but when it comes to riding the gravity train, those same scientists are passengers who know how fast their train is going and when it will arrive at its destination, but they absolutely cannot tell you what makes the train move.

It helps to keep our ego tightly fastened between our ears.

One last item to keep egos battened down. Man has a natural instinct to know whatever he can about the world and the forces about him. He's always searching for absolutes, what we might call the final answer. Getting to the top of the hill of any mystery is an overwhelming compulsion. Too often, however, we exhaust to the full the ability we have to look deeper into a problem and announce jubilantly that we have learned all there is to know on the subject. (You'd be surprised how many times there have been official recommendations to close down the U.S. Patent Office because "there's nothing left to invent." That sort of recommendation was made even *before* the Wright brothers made their first flight . . .)

Less than forty years ago science arrived at the end of the road in the study of the atomic structure of all matter. It was painfully obvious. The atom was the smallest particle that could possibly exist. Nothing could be smaller than the atom because the atom was indivisible.

To split the atom was *impossible*.

Well, at one time it was impossible to travel faster than thirty miles an hour. And surely it was impossible to travel around the world because you would go tumbling off its edge and break your neck (and be chewed to little bits by whatever monsters lurked beyond). And everyone *knew* that man could never fly.

Some people never learn that you can't do the impossible. So they went right ahead in their ignorance and they split the atom into a nucleus of neutrons and protons, with the nucleus surrounded by a swirling cloud of mysterious particles called electrons. That little act encouraged others to attempt the impossible, and they split a special kind of atom—uranium 235—and suddenly we had at our fingertips the awesome energy of the stars themselves.

Would you believe that at *this* point it was determined that the subatomic particles of the neutron, proton and electron were now declared to be absolutely the ultimate in the world of the atom divided? You can anticipate the rest. Suddenly new particles were discovered. And then even more. And as we built new machines and peered deeper into the marvelous microcosm, it began to dawn on us that suddenly we had found an incredible number of enigmatic particles and forces. There was not only matter but *anti*matter, a universe where everything was the absolute opposite of the universe we know.

Well, if this was so, calculated the stunned scientists, then there must exist a particle (which is likely a poor choice of words to describe something no one has ever seen) which they called the neutrino. Whatever had previously been mysterious paled before this baffling element of nature. For the neutrino, without measurable mass or charge, seems almost to deny the existence of matter as we have long thought of physical objects.

In the subatomic world, invisible to our senses, and often

THE SPLENDID DESTINY

existing only as expressions of mathematical formulas, we were being forced to deal with an unknown that defied every concept of common sense. The neutrino seemed to regard physical matter as no more substantial than thought. Any particle whatever, such as the proton or the electron, no matter how powerful, can be stopped by a wall of matter.

That's in our everyday, sensible world. To the neutrino, well, look at it this way. Light travels with a velocity of some 186,300 miles per second. That's about 676,000,000 miles per hour.

Sixteen billion miles a day.

Imagine, if you will, hurtling along at this speed every single day of one year. For three hundred and sixty-five days you clip right along with a speed of sixteen billion miles every day.

But you don't stop here. You travel at this same speed, never slowing down for a single instant, for fifty years.

A neutrino can move at the same speed through much more resistant matter such as lead, through which it could travel with only the barest effect, or none at all, upon its passage.

On that scale of energy the Earth doesn't even exist.

No two ways about it, this sort of thinking leads quickly to headaches. You begin to wonder what's real and what isn't. Albert Einstein was as baffled as you or I, and he was led to make a statement that has become the touchstone for all men of curiosity and wonder who would know more of life and its miracles.

> *Insofar as mathematics applies to reality, it is not certain, and so far as mathematics is certain, it does not apply to reality.*

But the one all-encompassing force that does apply is the mind of man, for it is here where all great voyages are born, and it was here—when the first man looked up at the moon and felt the urge to bridge the awesome gap between worlds—that the conquest began.

PLANETFALL

Now we have looked back from far in space and observed Earth in all its magnificent beauty.

We have walked the surface of two worlds.

We are preparing to leave man's footprints on all the planets of this solar system.

Forty thousand years from Cro-Magnon to Kitty Hawk.

Less than one generation of man from Kitty Hawk to the Sea of Tranquillity.

Can anyone truly doubt our journey to the planets?

CHAPTER 6

Prelude

This book of the future is also a story of the immediate past, and, in the chapters to come, will take us on a great and stirring journey of realistic imagination through the solar system. There will be more reality than would have been possible only a few short years past, for we are now veterans of sailing the seas of space. There is a strong urge on my part to convey to the reader that these words you read are in many respects deeply personal, and that descriptions of future events are based as firmly as they might be on a direct, personal involvement with the past.

It occurred to me as I began to write the chapter that would describe a future flight to the moon, when such voyages would be taking place on a scheduled basis, that I had been both participant in and witness to the most exciting times of all. I had been there at old Cape Canaveral when the first rocket ever launched from the Cape bellowed hoarsely to begin a new age, and through the twenty-three years since then, to this moment, I have lived through a period of time that will occupy future historians for uncounted years to come.

The story through all the years, from the moment Bumper burst into roaring life, is far too lengthy to tell in full detail, but there are selected moments that carry us along from then to now, and fill those years with a personal rather than detached view of all that has happened.

Make no mistake about it, these words come from no reference document in some musty library. They describe real moments that were experienced and felt very deeply, at times with a mixture of awe and even reverence. Let no one tell you, ever, that a man is not aware when he stands at the crossroads of the history of his race. That awareness fills the world about you, and these are the times when the flickering away of last seconds on the countdown digital clock is like feeling the heartbeat of history itself.

How many rockets have there been? As I looked back it became clear that every new thundering adventure from the Cape, and then from the Moonport, the Kennedy Space Center, and also from distant launching grounds like White Sands, Vandenberg, Holloman, Wallops Island; from Muroc Dry Lake; at isolated bastions where the great machines were chained to rock and then lashed into howling flame . . . Well, they began finally to flow together, to join in a kaleidoscope of fire and screaming thunder and shock waves and sorrow and exultation that merged details into that deep-seated *knowing*.

Long before the first boosters gathered their fiery energy for the crashing ride into the waiting vacuum so high above us, there had been rockets, missiles, hybrids, test vehicles, piloted machines, all of them sweeping together from success and failure the ingredients from which we would build the true giants standing in the wings, waiting for their time on the launch stands that would send them to other worlds. And beyond, to the first epochal journey of faithful robots away from this solar system, to drift on a forever voyage to another star. They are on their way *now*.

In the early days of the Cape there were no great rocket boosters, no satellites, no spacecraft. Nor were any intended, except among small select groups who saw in the burgeoning power across the palmetto scrub and sand the first glimmerings of the immense energy to be chained in some near future. Those were the days of Bumper and Lark, Matador and Snark; there was Bomarc and Pershing and Navaho and Mace, Redstone and Polaris. The missiles grew in size and power and performance, and the list of names was added to with Poseidon and Jupiter and Thor; there was Atlas and Titan I and the biggest missile of them all, Titan II. But across the country there were others, all of them contributing in different ways to the swiftly expanding horizon of knowledge and capability. We flew Skybolt and Hound Dog, Lacrosse and Aerobee, Corporal and Sergeant. There was Shrike, Rascal, Honest John, Lance, the Nike series and Spartan and many more.

While the hurtling messengers of war progressed from frequent exploding disaster to increasing reliability, men were trying their wings by building machines that were compromises of rocket booster and aircraft. It was a case of going carefully, treading along and then through the sound barrier, learning first to survive incredible shock waves, to traverse the invisible steel-hard reefs and treacherous shoals standing between the air ocean and the waiting vacuum beyond. One by one they came on the scene, powerful little machines known as XS-1 and X-1A and D-558-II and X-2 and finally the X-15, a huge black brute that smashed its way upward out of atmosphere and took its men so high that outside their sealed cabins the air was but a millionth of the density at sea level, and when they returned again to Earth they had earned the honor of wearing astronaut wings.

It happened together, a journey clawed from frustration and dying men and an unshakable belief in where we were going, what we must do; it was a vast effort, really, most of the work and the tests running concurrently. From this vast orchestration the decision to use the machines and the skills emerged almost hesitatingly. Those who were in a position to comprehend the urgent need for placing instruments and men beyond the atmosphere were in no position to order such things done, and those who had the political power lacked understanding. Rockets were regarded as great weapons, but finally those in the higher levels of government were convinced that the same rocket that could send a massive warhead from one continent to another at 15,000

PLANETFALL

miles an hour could be modified into a rocket that would send a hundred, and then a thousand, and then ten thousand pounds of scientific instruments into orbit high above the world and, later, far beyond Earth to other worlds.

That was the bridge to be crossed, and it was no more than that. The beginning of what we know as the age of space required not a single scientific breakthrough, not a single engineering miracle. It had all been done in the ominous rockets built for global war. Early in the 1950s, both in the United States and in the Soviet Union, men of vastly different political faiths came to the same conclusion advanced by the scientists of each nation. From that moment on it was simply a matter of time, and on October 4, 1957, the shot was fired that was heard, and *seen,* around the world. Sputnik I raced into orbit at nearly five miles a second, and the world would never again be the same.

At Cape Canaveral—soon to become Cape Kennedy—new names emerged from the transformation of missile to booster. Vanguard, Jupiter-C and Mercury-Redstone, Juno II, Thor-Able and Thor-Ablestar and Thor-Agena and Thor-Delta, Scout, Atlas and Atlas-Able and Atlas-Agena and finally Atlas-Centaur. The Titan I missile developed into the Titan II and then the IIIA and finally into the massive IIIC, and, in the wings, waiting to launch huge Viking payloads to Mars, is Titan-Centaur. The Saturn I and IB were birthed at the Cape, before IB moved to the sprawling Moonport we know as the Kennedy Space Center. And finally, the supreme giant of them all, the Saturn V, the monster to hurl 240,000 pounds in one incredible firing into Earth orbit, and whip more than 100,000 pounds to the moon.

The scientific probes gained a galaxy of names that began with Explorer and Vanguard and Pioneer and Score, and then expanded quickly with Tiros, Orbiter, Injun, Surveyor, Beacon, Discoverer, Transit, Midas, Samos, Relay, Anna, Starad, Ranger, Alouette, Telstar, Ariel, OSO, OGO, OAO, Mariner, Vela, Syncom, San Marco, Nimbus, Echo, OV, LES, Surcal, Secor, Peg-

PRELUDE

asus, LCS, Early Bird, Snapshot, GGSE, Biosatellite, ATS, Intelsat, ESSA, Scout, Pageos, TETR, WRESAT, Tacsat, Aurorae, HEOS, Anik, Iris, Skynet, Azur, ISIS and Boreas—a list that is far from complete.

But of all the pillars of fire that stood above the Cape and the Moonport, the names that would ring through all the world were Mercury, Gemini, Apollo and Skylab. They will be engraved forever with Vostok, Voskhod, Soyuz and Salyut as the machines that were first to be launched on that incredible ocean through which move all the stars and their worlds.

Looking back on the twenty-three years of nearly two thousand launchings from the Cape and the Moonport, certain moments stand out much more than others. It is strange to realize that the most *impressive* flights are not necessarily those marking the greatest risks or those that were the most important. One launching may, of course, be representative of many others. And after being witness—sometimes standing no more than a hundred feet away from some of the missile tests—to so many overwhelming launches, those that had the greatest impact on the senses are most strongly brought to mind.

Early in 1955 I spent a great deal of time at the Air Force Missile Test Center (now the Air Force Eastern Test Range), working on several highly secret projects, one of which was a proposal for our first attempt to send a payload to the moon. Dr. Wernher von Braun, Colonel Don Burrus and I did considerable work on this project which, if never making it off the launch pad, at least started the momentum that would, three years later, see the countdown for the actual firing. On the morning of April 15, 1955, I was at the Cape, waiting for six hours for the first Redstone missile to be fired at night. It was ten minutes before two o'clock in the morning when the searchlights at the launch pad snapped out and the countdown rushed to its moment of truth. The sight was so overwhelming—keep in mind that this was the *first* large rocket we had ever launched at night

PLANETFALL

—that I wrote down the scene before dawn of that same day. These are the words as they were written:

A light flared . . . a golden glow that spread rapidly around the base of the missile and illuminated the launching area. The soft glow lasted perhaps a second; then the bird lifted and behind it we could see an intense, deep, rich, yellow-gold flame, brilliant and sharp to the eye. She was up! Not a sound yet, but she was lifting faster and faster.

And then, when she was well off the ground and the light was racing over the Earth, like the spread of dawn climbing into the sky with the bird, we heard it for the first time. In all my life I have never experienced such a sound. It wasn't a noise you usually hear; I've seen enough launchings and heard enough to know that at once. This time it was different. The rocket was high now, several miles up, and the sound came from all over the heavens. It sounded as if someone were playing a mighty celestial organ, and they had jammed down on the bass pedal. It was pure sound, without a break, without a flutter. One steady, deep, roaring beat, unlike anything I have ever heard before.

Imagine this incredible celestial thunder, all the while the rocket climbing higher and higher, going faster and faster. For a few seconds the picture was fantastic, unreal. It was hard to believe that I was watching a scene from real life instead of looking at some cleverly created special effects on a three-dimensional movie screen. Redstone was several miles high, and her light spread in white and gold from one end of the horizon to the other, like a muted sun that had leaped suddenly into being. Then the horizons lapsed again into darkness, and the shadow raced back until the entire world again was dark, except for that incredible searing point of light now leaping into the stratosphere.

My head was bent way back, watching the rocket ascend. The light was only a pinpoint now. Redstone was almost twenty-three miles up, still flaming along on her engine. Then the rocket flame cut off. She was still coasting up, beyond what we know as atmosphere. We could still see her; the vanes in the exhaust glowed enough to be seen at more than twenty-eight miles. Little

PRELUDE

by little they dulled. I strained to see . . . the glowing red was in a field of millions of stars. She was leaping away from Earth, higher and higher.

Right then and there I knew I had seen a preview of what the moon ship would be like when she left the Earth behind.

That was the first. On August 17, 1958, the first Thor-Able booster of Project Mona tore away at dawn from its launch stand. Seventy-seven seconds later the morning sky high over the Cape filled with tumbling, flaming debris as the first attempt to reach another world ended in a heart-rending explosion. The second attempt also failed, and then, on October 11, 1958, we were ready for the third, a night launch preceded by a dazzling display of lights at the launch pad that sent a fantastic crisscross of glowing beams into the sky.

The night was cool, with a mist hugging the ground, pressing gently like some milky liquid about the scrub pine, the palmetto, the countless poles and towers of the Cape. From the side of the rocket supercold gaseous oxygen vented in a billowing vapor that, caught by a backdrop of lights, became a mass pirouette of swirling incandescence, richened throughout with a predominance of violet. The vapors became creatures at the whim of the slightest breath of air, always in motion, and from the mass of ground-hugging vapor there flowed wispy tendrils, luminescent creatures curling softly in a mystical fairyland.

The count rushed to zero, and within the moon probe electrical and hydraulic pressure raced into life. Small vernier engines on each side of the booster shrieked into life, stabbing knife blades of flame sideward and downward. From a mile away the moment of ignition was blinding and exquisite. A searing blast of orange exploded in absolute silence; instantly the mist across the Cape breathed in the orange glow. It was a miraculous transformation, light that was soft at the far edges of the Cape, brighter toward the launch pad, obliterated near the rocket where flame slashed viciously against steel. Clouds of

PLANETFALL

steam boiled away from where flame met high-pressure jets of water. The rocket seemed abruptly to break its bonds, and it rushed upward from its hold-down ring: a perfect, flawless, stunning ascent. Then the sound crashed outward all across the Cape. She rushed away from the planet of her birth, the thunder increasing, a sonic blowtorch crying its song of power, the golden light rising majestically.

Night became orange day with a shock. Hundreds of small animals scurried to cover from the unexpected splash of an orange sun; thousands of birds whirred in startled flight from the ground. The light dwindled then, but the roar hung suspended between heaven and Earth.

Through powerful binoculars I watched the main booster engine wink out. Then, holding my breath, watching only the backdrop of stars, I waited for a splash of fire along the edge of space, the ignition of the second stage. Then . . . there was no mistaking the sudden blaze of light dimming the stars. Ninety-two miles above the world there appeared the tremulous flame. What happened next was totally unexpected. Fire washed back, raging, against the Thor booster that had been discarded. It splattered wildly against metal and gave birth to an unprecedented play of light we had never seen.

Globules of flame, fiery teardrops in all colors, stunning . . . A soundless explosion ghosted through space, with a million sparkling lights, as the flame struck the Thor and scattered in vacuum. For a moment, clearer than I had expected, I saw the body of the long rocket slew sideways, fire gleaming all along its white body. The light vanished, and night reclaimed its domain.

There were, of course, many brief moments when each incredible sight seemed to transcend all others. But each was a moment of glory and awe and sometimes frightening in its own special way. Nothing could compare with the sight of the Cape as seen from thousands of feet over the ground at night, when a rocket tearing into fire on its pad seemed to set the entire world

PRELUDE

aflame before it rose, drawing in the light behind its ascent as though it had given, and was now taking away, a light of pure gold.

I watched a sight one night that no one expected. That night I was the pilot of an Aztec at 14,000 feet over the edge of Cape Canaveral. Four newsmen with me waited for a Thor-Delta to emerge from storm clouds only three thousand feet below us. We listened to the countdown on radio, ticked off the seconds aloud as the rocket raced upward, invisible to us, through the clouds below. And then we knew, with a sudden and *very* real fear, that we were much too close to the launch pad, its exact distance obscured by the storm below. The clouds suddenly glowed silver. We had never expected that. Gold, yellow, orange, or red, but not silver. No matter. Silver it was, a billion sparkling motes spreading swiftly and silently from a single dazzling sun shredding the clouds and imparting firelight to the rain in all directions. The rocket erupted from the storm, and, it seemed, it hurtled directly at us. There was frantic movement as I stood the plane on its wing, full power to the engines, trying to get out of the way. We never knew how close it came, but we could hear the roar, the thunder louder than our own engines, and it seemed as if a silvery sun was hurtling upward alongside us.

We were still shaking when we eased down through the clouds and returned to the airport.

One of the more memorable moments came *after* the rocket had left its launch pad and disappeared from sight. Apollo XII, my three good friends Pete Conrad, Al Bean and Dick Gordon at its peak, launched in a pouring rainstorm. They seemed to lunge upward in the heavy rain, the clouds swallowed the mighty Saturn V rocket, and only the thunder smashed back at us as the Saturn hammered her way aloft. But the rocket, 363 feet in height, and a tail of fire another 500 feet, became a sort of super lightning rod. Thirty-six seconds after lift-off, as we still watched the steaming launch pad, jagged bolts of bluish lightning curved down from the rain-drenched sky and smashed into the ground

PLANETFALL

and the launch pad—and we knew that Apollo XII had been struck by several enormous bolts of lightning.

There's no doubt of it. We were more frightened for those three cool men in that spacecraft than they were for themselves. At least they were busy getting things back in order. All we could do was stand and wait to hear from mission control, because all that could be seen of the world was rain. It was the longest single period of time I have ever known at the Cape or the Moonport.

There are but two other moments with which to bring this chapter to its close—each was a last flight, and each took place by night, and each, in its own way, proved to be visual paintings the like of which had never been seen before and may never in the future know their equal.

The first launch took place on January 29, 1962, while we were sweating out the interminable delays in getting John Glenn off the pad for the first orbital flight of Project Mercury. The Air Force had planned a late afternoon launch of the last Titan I in that test program, but there were so many delays in the count that by the time the final seconds ticked away the sun was just below the horizon. We were, then, in final twilight. High above us it was still day. On the ground, darkness fell quickly, and brilliant searchlights came on at the pad to illuminate the thick-bodied rocket.

For this mission I had managed, with the aid of an Air Force major who had no more sense than I did, to work our way, unseen, to a point on the service road that was barely more than a thousand feet from the launch stand. We stayed hidden until the final seconds when we stepped out onto the road for a senses-overwhelming view that was, I'm sure, like looking directly down the throat of a raging volcano.

The searchlights flashed on the gleaming ice shrouding the rocket's tanks, bringing her to loom huge and powerful, casting off streamers of bluish-white gox (gaseous oxygen). At our distance we could hear the shrill scream of supercold fuel in its pipes, a shrieking chorus set to its outlandish cry by the mixing effects of temperature and wind. Darkness was all about us now and the first stars appeared, and then we were in the final fading time-swell of the countdown.

The searchlights winked out almost as if darkness had stamped out the glowing beams. The squawk box set by the roadside chanted the count and I watched the vapor plumes disappear as the liquid oxygen tanks were sealed for pressure build-up. I put aside my binoculars for the moment. At this distance they would completely blind me.

At thirty-one minutes past six o'clock the launching pad, barely more than a thousand feet away, exploded in a glorious burst of golden fire. Light flashed in all directions as Titan broke her gravity chains and began her ride away from Earth. Then it was impossible to see anything but light—savage, intolerable light of pure gold from the twin-engine chambers. A thousand feet distant? Impossible! They seemed to be only yards down that road, pouring forth light so intense it bleached, instantly, the last drop of blackness from the night.

The golden fireball lifted silently, then, without warning, hurled forth its sound with body-punching fury. It seemed that a great knife of sound had split the sky asunder. It was no longer sound but a volcano erupting shock waves. You no longer *heard* the sound; it punched and embraced and squeezed the body, and everything in the world vibrated in resonance with the *thing*. No longer could I see the rocket, only the white outline of the curving steel flanks hidden by the golden fire. The thunder brushed aside its harsh overtones, and the wave of sound receded to the familiar acetylene-torchlike howl of a giant running mad.

Faster and faster she rose with unstoppable acceleration. The flame revealed thin hollow tubes of fire, within which shock diamonds paraded in an endless stream. I placed the powerful binoculars to my eyes; the distance shrunk magically and I was seeing the wonder—close up—of Titan racing away. She bent in her path, leaning into the southeast, and the yellow became deep red.

PLANETFALL

Suddenly Titan was free of the night. It was still evening and the line of shadow cast by the rounded bulk of Earth stretched far over our heads. Titan had reached above this line and hurtled back from night into day. Abruptly the long rocket and the fire trail were struck by the sun and disappeared beyond our horizon on the ground. First the flame became a blood color, then, with increasing height, a rich orange. Titan had reached the edge of space.

And now began the most beautiful visual fantasy ever known in this age of space, what the awe-struck humans standing so far below have named *Aurora Titanalis*.

Behind the flame appeared the contrail—the long, wide swatch of condensation vapor. For a moment it was dark; struck by the sun's rays, it became a multicolored streak rising higher and higher above the Earth.

Titan exhausted the fuel in her first stage. In near vacuum the two blazing chambers shut down. There came a moment without fire, but I could still see the deep red glow of the chambers. Suddenly the upper-stage engine exploded into life.

And then began six hundred seconds of sorcery.

In the silence of space a glistening halo appeared around the burning upper stage. Ionized gases from the accelerating rocket rushed about in all directions, spreading the halo with tremendous speed. Normally it is invisible to the eye, but at this moment, so high, it absorbed the full illumination of the sun, while we remained in darkness.

The halo spread wider and wider, a huge pulsating globe that became a pale green in color and then began to reveal glittering spots of pink and deep purple. The teardrop expanded swiftly until it was more than a hundred miles wide.

In the midst of the intensely glowing gases in space, deepening steadily in greenish hue, lights appeared and began to move. The second stage of Titan burned fiercely, but through the plasma teardrop the color was not the familiar red or yellow or orange; it had become a white-green with pulsating sapphire points. As

PRELUDE

it rushed higher and faster with every passing second, it produced its own bulging gaseous envelope, a growing circle within the much vaster plasma teardrop.

Soon I watched the light coming back at us in the form of whorls and eddies, great glittering spirals, one after the other, a plasma whirlpool in the heavens far above the upper edges of the atmosphere. Still the Titan engine blazed, pushing the rocket higher and higher.

Then, to the right and below this light, appeared a single blazing jewel, a hard, white-green light *not* from Titan. This was the planet Venus, gleaming through the swirling kaleidoscope of green.

Still another light appeared! Higher than the rest, it seemed, but this was an optical illusion. The empty first stage of Titan, tumbling end over end from space, dropping back into the atmosphere. Friction tore at the metal skin; the steel glowed, then began to burn, and the twisting, tumbling fall of the flaming rocket added an orange-purple sheen to the sight before us. Without the binoculars, the sight was incredibly huge, commanding the sky. The teardrop had stretched, wider and longer, and now it became an enormous white balloon of wispy substance tinged with the green hue of the swirling eddies.

Through the binoculars, however, the colors leaped into prominence. The pulsating rocket engine kept the strange whirlpool of space spinning madly, flinging its glittering ribbons in all directions until, in the minutes that passed, I was seeing through several layers of whorls, all moving, all clashing and passing into and through one another. It was the most fantastic, beautiful sight I have ever seen, and I have yet, to this day, to see anything remotely like it.

The first stage appeared to be higher than both the still-burning upper stage and Venus because of the Earth's curvature. The upper stage was bending far over in its flight, beginning to follow the arc of the planet. But the first stage had soared up and over, and was now coming down, much closer to us. And so it *seemed* to be higher—just as a person looking at a low-horizon moon, from beneath a tree, may "see" the moon as "beneath" its branches.

Slowly, achingly, the celestial display played itself out to a ghostly wisp of the awesome sight which we had been granted. Finally, only the gleam of Venus remained to pierce sharply the dissipating teardrop in the heavens.

And, finally, there is the launch that cannot be described—the last Apollo. Only a part of that most awesome of all moments at the Moonport may be put to paper. Only the early phase of ignition, the lift-off, the climb away from the soaring tower. That's all. When Apollo XVII touched a place in the sky perhaps two miles high it transcended words. It was not something you saw. It was an experience you lived, and you were overwhelmed and cowed and struck down with the knowledge, right then and there, that you were witness to a moment in time that truly heralded the race of man achieving a stature and a glory our ancestors could never have begun to comprehend.

Let me stress that there is nothing in the world—*nothing*—that compares to the Saturn V. Except for the Russian boosters, I have seen every big bird of this country smash the Earth with fire and pound spaceward, and the biggest booster of the Russians—the Proton—is something on the scale of the Saturn IB and the Titan IIIC, which is a *long* way from the Saturn V.

The huge launch vehicle for our Apollo program, and the Skylab space station, doesn't even sound like a rocket. At least it doesn't sound like you think a rocket is going to sound. Every Saturn V reminds you that the rule book goes out the nearest window at ignition. Unless you've been there you have never *heard* the Saturn V. Not on television or radio or on tape, because the sound is so overwhelming, so filled with smashing concussion, that it overwhelms any recording device or system ever built. The huge booster doesn't send out the roar to which we had become so accustomed through the years. The sound is a

PRELUDE

roar *plus* shock waves that are visible in the air *plus* an overtone of explosive crashes *plus* a series of hammer blows, repeated swiftly one atop the other, that literally, and I mean literally, vibrate the air and shake people and buildings and set the Earth underfoot rumbling. What I've just managed to do is tell you what the sound does and everything it *isn't*. But there's no other way.

The Saturn V lifting off is not a launch. It's an experience. The most callous observer, the most disinterested witness who comes to the Moonport for prestige or curiosity, or whatever, always leaves shaken to his core, perhaps not understanding what happened, but *shaken*. You don't just watch and listen to the Saturn V, you are swept up and squeezed in a monstrous fist and when *it* gets ready you're left weak and wrung out and the word humility drifts into your mind, and when you try to force yourself to realize, to understand that three frail human beings are riding that incredible vertical mountain out there, and that they'll be howling toward the moon at 25,000 miles an hour . . . it all crowds into your head and rattles you right down to your heels.

Well, all that had happened with all the Saturn V launches, all of them, just went away when Apollo XVII *happened*. This was the last flight of a manned Saturn V, the last flight of Apollo, the end of the beginning of the manned exploration of the moon, and the first and only night launch of a rocket that weighed nearly six and a half million pounds standing on its mighty launch pedestals. Enough of that, enough of the introduction, and let's get right to that instant when the mountain exhaled its first shocking gasp of flame.

The rocket stands alone, utterly removed from all life save the men strapped into their couches atop the mountain. There it is, waiting. Then, appearing from nowhere, a great ball of flame, an incredible red-and-black mottled rose exists beneath the five gaping engine chambers. The rocket is alive, but this is only its first weak sign of life. The fuel is not yet being rammed under full pressure to the engines, but when it is——

PLANETFALL

For a moment you watch the boiling, twisting rose, shredded to torn, flying petals as the cataract of water to cool the launch stand tears at this first fire. Only that moment. Then quickly, as fast as your eye can follow, the engines are fully alive and there is *flame* like none other ever created by man. It is a volcanic birth in brilliance, in size, in fury, a single heart-stopping outpouring from hell. Now, faster than you can see it happen, the flame rips downward, into the curving bucket that bends the fire on each side of the pad, and shrieks left and right of the Saturn.

What happens now is that the light spreads and spreads and becomes brighter, and to each side of the rocket blinding flame and glowing steam rise for hundreds of feet into the air, boiling and twisting upward. For nine seconds that incredible monster stands where it is and screams its defiance at the world, continuing to send its savage flame into the sloping bucket cavities to be embraced by the Niagara of cooling water. Still it doesn't move as it builds up even *more* thrust, more energy, and it shakes through its whole vast body, sending great chunks of ice, glittering sheets of snow from its flanks into the fire creature below. It is a snowstorm in hell and there is no other word for it.

Ten massive steel arms snap back, away from the thick tail fins; the servicing walkways are dragged by pistonlike pressure away from the sides and there is no argument—the raging fire giant is free.

And you're holding your breath because she moves slowly, so slowly, and you can't see her because as the flame shatters downward, free now of the mass of steel and concrete that concealed most of it from view, it is all too bright, too blinding, and you are watching the birth of a small sun, a continuing, impossible, brain-squeezing atomic fireball pushing away from Earth. Still everything is in silence, for it takes eighteen seconds for that thunder to rush and pound and sweep its way from the launch stand to where you are. In that unbelievable silence your eyes adjust as you squint, and you can make out, barely, the details of that blinding mass of golden fire, intense, whipping throughout

PRELUDE

its substance, shimmering along its edges, trailing into ghostly purple flame far below the now climbing giant.

Only that brief moment was given for Apollo XVII; the rocket was gone forever in its concealing fire that tore at the eye sockets. All this had continued in silence, and *now* the storm was upon us, and it hit with all the force of a truck, the shock waves colliding and banging and slamming and crashing together, a thunder so loud it tore itself to pieces and kept bouncing and rebounding within itself, a ragged chorus of overwhelming, swiftly-repeated staccato thunder to make a crying child of the mightiest thunderstorm. Each blast, tumbling and crashing, each only a part of a second, a powerful blow unto itself miles away. It was painful to the ears and it was so dominating you stood frozen where you were, feet rooted *into* the ground, and you were slave before this ultimate master, and you knew you were hearing the glorious and triumphant cry of the giant smashing the chains of gravity, casting away the fetters of two million years. There it was, the billion-throated dragon tasting sweet escape, hammering upward in its assault to rush into the depths of vacuum where it would live to raging, splendid fury its brief life, to have served well three men destined ot sail to another world.

And this was only the beginning. But there is no way to describe the rest. Apollo XVII was perhaps eight thousand feet up now, and all that may be said is what a man felt. For the entire sky was burning. That is the word for it. The clouds and the sky were burning, consumed by this ultimate of all fires, and there was no night or even semblance of night, there was only an entire sky sundered by flame. The sound had increased and was now a thing berserk, and it all came together, and we stood there in awe, many of us with tears on our cheeks, because it was all too much, too much to absorb and——

If ever we may be privileged to see God drawing all his angels to him and ascending into whatever heaven there may be, and all the world is filled with blazing glory and the sky burns . . . it will be like that.

CHAPTER 7

Outbound

If we could turn the clock forward . . .
Well, why not? Isn't that just what we have done before every new exploration, before every new adventure? We gather in one place every last scrap of information and do our best to extrapolate—which is educated guesswork—as to conditions we'll meet when we get "out there." When we first sent robots to the moon and followed with the manned flights of Apollo, we expected two things. First, much of what we already knew about the moon would be true. And so it happened. The surface of the moon was a vacuum. That surface was lashed with sweeping radiations (against which their pressure suits fully protected the astronauts). On the surface of the moon there would be only one-sixth the gravity we experience here on the surface of Earth, and a 180-pound man weighted down with 180 pounds of suit and equipment (that's Earth pounds, remember) would weigh only sixty pounds on the moon. Getting around would prove to be a bit clumsy because of the restrictions of the suit, but it would also be a delight. And it was.

We were wrong about just as many things as we had been right. Those jagged mountains and sharp rocks didn't exist. Places where we expected to find volcanic material often turned out to be puzzling disappointments, and where there weren't supposed to be lava flows, the surface was filled with lava. The layer of dust on the surface was comfortably thin, but it was a kind of dust that worked its way into equipment and began to grind down and jam up so many vital parts that it became dangerous.

Well, that's the way it went, winning some and losing some, but in the really critical departments—basic characteristics of the moon, the performance of the Apollo machinery, the marvelous adaptability of our astronauts, to name only a few—we were right on target. We could gamble men's lives on forecasts based upon as much information as it was possible to accumulate and know there was every chance those men would return safely.

Now we look with the same eagerness, but with even more confidence, to our explorations in the near future. Imaginative probing of the years to come is absolutely necessary as the first step for going from Here to There. *Planetfall* is an exercise in both hard fact and careful prediction, and we enjoy the option of making it a scientific adventure as well.

We know about getting to the moon *now,* because the history of Project Apollo is so recent that the only dust on its records was that brought back from the moon. Between Christmas of 1968 and the end of 1972 a total of twenty-seven American astronauts orbited the moon, and six of the twelve men who walked that alien surface had the pleasure of driving across its dusty rock.

One reason why we may commit to what we call hardware predictions of future activities away from the Earth is this experience. So many thousands of hours have been accumulated away from the home world of men that we now enjoy a solid foundation of anticipating situations. We have seen our capability to sail the seas of space emerge from pure speculation and, in an astonishingly short period of time, become stunning reality. Perhaps that is the first lesson to heed: no matter how "impossible" accomplishments in the future may seem, progress has the delightful habit of exceeding our fondest dreams.

In 1950, before Cape Canaveral was host to its first crude rocket, the best, the most expert, the most knowledgeable authorities in the field almost unanimously stated that man would not reach the moon before the turn of the century. In 1950 the moon

PLANETFALL

could not possibly know the booted feet of man for another fifty years, that is, until the year 2,000.

That was 1950. Five years later the eminent astronomer royal of England reiterated this conviction. No, he went further than that. In response to stories that man would one day be able to journey through space, this leading scientist snapped that space flight was "utter bilge."

Two years later the first satellite went into orbit about the Earth.

Four years after that opening shot in the age of space, men orbited the planet.

Seven years after the first orbital flight, three men looked down on the craters and mountains of the moon and, as they orbited that lifeless world, sent by radio and television to the awed planet a quarter of a million miles away Christmas greetings to the people of the "good Earth."

Seven months later two men were on the moon.

Eleven years after the first man raced into orbit about Earth the first wave of manned lunar exploration was *over*.

One of the lessons of history is that events that seem impossible, or beyond realization in any future within sight, rush upon us with little or no warning. Almost inevitably we are super-conservative about the speed with which we rush into the future.

In 1918, World War I—the war to end all wars—came to its end, with an appalling toll of human suffering and loss of life. The concept of another world war lay beyond the imagination of sane men. Nineteen years later, World War II began with full-scale war in Spain and China.

In the next decade alone the world was introduced to the jet age, dazzling developments in medicine, the release of atomic energy, the impact of electronics, rockets that reached into empty space above the atmosphere and the computer.

All in just about *ten* years.

So we will make a great leap forward to the year 1998 for our travels through the solar system. Just about a quarter of a century. That gives us plenty of leeway. After all, we've had manned space flight for only *half* that time.

And things do happen much faster than at any time in our history.

It took Marco Polo twenty-four years to make his round trip from Venice to China and back home again.

It took the first Jupiter probe eleven *hours* to race past the moon after launch from Cape Kennedy.

The year is 1998 and we're about to depart Earth, outbound for a manned flight to the planet Venus.

With a stopover around the moon.

Once upon a future time . . .

Cape Kennedy is a museum. So is the sprawling Moonport nearby. The last huge rocket to thunder upward from the Florida coastline eased into orbit in 1987. This was the famous utility spacecraft that followed the towering Saturn V boosters of Apollo, the winged Shuttle. From 1979 until 1987 the Shuttle was the backbone of most space flight activities of the world, used almost as much by European, Asian and South American countries as it was by the United States. The Shuttle changed manned space flight from a repetition of nail-biting adventures to a planned schedule of missions. But it was still wasteful in terms of energy. Two enormous solid-propellant rockets had to be expended every time the Shuttle flew. So did a great tank from which the Shuttle drew a small lake of fuel to get into orbit. The aerospace ship itself, the Shuttle, about the size of a small twin-jet airliner of the 1970s, was the only part that returned to Earth to be sent back into space.

Yet it was an overwhelming advance against the crude rockets that preceded the Shuttle flights. Within a year after the Shuttle was operational, every other launch pad at the Cape and the Moonport had been closed down. Shuttles roared into space from the Moonport and from Vandenberg Air Force Base in California. They took men and materials into high orbit to build the huge Spacelab station that followed the older Skylab launched in the spring of 1973. It was, despite the steadiness and reliability of the Shuttle that eliminated much of the danger of the older flights, still an exciting period. The world came to accept industrial, medical, scientific and engineering experiments and processes in space just as naturally as it did those activities around the surface of the world.

But still the Shuttle had its problems. It used chemical fuels in such abundance that ecologists turned all their pressure in Washington against Shuttle flights, claiming the rocket exhausts spread dangerous chemicals into the atmosphere. The truth was far from the claims, but the entire world was recoiling from the stink of pollution, and public pressure to find a better way mounted steadily.

Scientists were in full agreement, but for different reasons. Chemical rockets are clumsy and inefficient. They're the lumbering barges of the space age. For a while the big promise for leaving Earth was the use of nuclear engines. With the dumping into the atmosphere of radioactive gases and the chance that a nuclear rocket could explode, *that* idea didn't last long. There was a better idea. It had been around since the early 1960's, but it demanded much more research.

The big breakthrough came in 1982. In the laboratory, anyway. Scientists for years had tried to improve the tight-beam broadcast of tremendous energy. Energy is broadcast in the form of radio, microwave and laser beams, of course, but those weren't the kinds of power levels the scientists wanted. In 1982, computer studies produced the results they had been waiting for. Suddenly we had the means to transmit on narrow beams—much like the microwave beams of old radar—enormous levels of energy for many miles.

Out in the Nevada desert, hidden from the world within the old atomic bomb testing grounds, the government launched its new energy beam-transmission project. If the new concept worked it would be the closest thing to antigravity. It wasn't anything of the sort, of course, but it was a quantum jump ahead of the clumsy chemical rockets. Engineers built enormous power grids across the floor of the desert. Huge nuclear reactors to generate vast flows of energy went up. From the edges of the grids, powerful beam transmitters went into place.

In its simplest explanation, the grids channeled the power from the nuclear reactors to the beam transmitters. These focused the power, drawing from the grids to create narrow beams that were transmitted to a spaceship standing on a high platform in the exact center of the beam transmitters. The one legacy that remained from the "old days" was the countdown. When a ship was almost ready for lift-off, the energy receivers were turned on. Huge hold-down clamps kept the ship on its stand as power built up. The energy receivers took in the microwave-transmitted power and channeled it to the repulsors built around the base of the ship.

When the electromagnetic thrust transmitted away from the repulsors, as the force would be measured in pounds, exceeded the weight of the ship as it stood on the Earth's surface, it began to rise. Slowly, majestically, with only a deep humming sound and the fierce glow of ionized air about the repulsors.

The greater the distance an object is from the surface of the Earth, the lesser is the planet's gravitational attraction. Therefore the object—a spaceship—"weighs" less. If the energy being used to lift that ship's mass from Earth remains constant (or nearly so), then the ship will accelerate constantly. There's no need to rush in the early stages of flight because the ship is also fighting the resistance of the atmosphere. But at twenty miles up it's above everything save tenuous wisps of air, and at this point the acceleration can increase swiftly.

The first tests of energy-beam transmission were a magnificent success. The government hurled billions into the program, which had a side benefit. When the huge nuclear plants weren't being used for lifting the great spacecraft away from the Earth, their power was used to provide electricity for hundreds of miles around. It was an eminently satisfying relationship in every respect.

One problem that still hasn't been overcome is that the energy beam has an effective range of just about 130 miles. A ship racing into orbit needs a speed of about 17,500 miles an hour for low orbit above the planet. But it doesn't climb straight up, of course. It bends over in its flight to more and more parallel the surface of the globe below.

The old nuclear rocket program of the 1960s was a vast sinkhole of frustration. Scientists had worked for years on the Nerva rocket, then had to abandon the program when the money spent wasn't justified by results. By the 1980s, however, nuclear propulsion had made its own vast strides, and nuclear engines had been perfected. No one bothered to build the same type of huge engines once needed for the old Saturn rockets. The power grids boosted a spacecraft to about 15,000 miles an hour before beam transmission lost its effectiveness. At that speed, and well beyond the atmosphere, the spaceship went to full power with its nuclear drive. Steadily, with amazing smoothness and quiet, the ship reached the speed it needed to go into orbit.

It was possible to build energy-beam spaceships of almost any size. But there was still the problem of getting back through the atmosphere, and no one has discovered how to work the energy transmitters in reverse. That's the big project under way now, of course. If they can increase the range, then enormous energy beams will flash out into space, lock onto a ship in orbit, and start deceleration. With these "tractor beams" operating, the spaceship would be drawn down along a massive invisible pillar of energy and brought to rest without a bump in the center of the grid power field.

But that's in the future. Right now the return to Earth takes place much as it did with the old Shuttle. The spacecraft fires a burst of energy to "deorbit." The speed decreases and eases the ship into the atmosphere just as in the old days. But today's metals and ceramics are so much more advanced that the ride, with the blazing streak of reentry, is exciting without being dangerous. Back in the lower atmosphere the spacecraft extends powerful fanjet engines into the air, and flies just like any other big airplane.

The combination of power transmission along tight beams, and the use of nuclear drive for operations in deep space only, has brought on a dazzling revolution. All of a sudden space flight was available at almost any time, as far as departing Earth went. It was so cheap to send huge payloads into space (because the electrical power used for industry and cities was paid for at commercial rates) that it was now possible to assemble in Earth orbit just about any size deep-space machine we wanted.

The nuclear drive had thousands of times more energy than anyone had dreamed was possible back in the days of Apollo and Skylab. That meant we weren't prisoners of energy conservation in moving through the solar system. The old way meant building up tremendous speed as fast as possible because of the limitations of chemical rockets. Then you coasted the rest of the way to your destination. That's why it took so many months to reach Mars or Venus, even when they were closest to Earth. Now a ship can accelerate steadily from orbit about the Earth or the moon and build up enormous velocity, since it also has the means to decelerate for long periods of time. This sort of spaceship drive proved an invaluable boon in yet another way. Not only did it reduce the transit time for the long space voyages, but while the spaceship was under acceleration or deceleration it created its own level of gravity and eliminated the problems of being weightless for excessive periods of time for the crews.

Within a few short years of the energy beams becoming operational, we had a fleet of ships that would never again touch down on the surface of any major world. To travel to and from the moon we—well, let's make the journey ourselves.

Who would ever have named a spaceport Jackass Flats? But that was the name some forgotten gold miner gave to the scorched desert in Nevada when the country was still isolated from the rest of civilization. It stuck, and Jackass Flats became a permanent part of our maps with the establishment of the elaborate atomic bomb testing facilities. It's still called by this name, although the countryside we see below us from our jetliner shows only a few traces of the wilderness that once was here. With almost unlimited power from the great atomic reactors of the spacefield, mass irrigation became feasible and the desert bloomed. The ground below is a riot of green, with long canals and blue lakes brought to life by the miracle of available energy.

We haven't much time to study the scenery, as our jet decelerates while it's still high in the air. The powerful engines hurl their thrust downward as well as behind us and we settle to the runway with a forward speed of barely sixty miles an hour. A slight bump and we're down.

At the spaceport, we're taken to a building that's about as close to a hotel as we could imagine. It *is* a hotel, but it's also a hospital and many other things. That night we must go through the mandatory physical examination before being permitted to board the *Evans,* the ship that will take us into orbit. The flight won't be punishing (that was for the old days, remember), but there will be some time when both crew and passengers will be weightless. That mandatory physical means an examination for the heart, the respiratory and circulation system. Just to be certain serious problems won't arise. As far as being sick from weightlessness, a mild drug takes care of that problem. Most people, however, after a short time in space, enjoy being weightless and actually look forward to it.

By eight the next morning it's time to board. There's no need for pressure garments—the spacesuit—anymore, but we are required to don a long-sleeve jumpsuit. So that any time we go weightless, this will keep things in our pockets from floating away from us. We strap into thick, yielding seats that are as much couch as a padded chair. The straps cover us across the waist, and there's also an inertial-reel shoulder harness. We have full freedom of movement with the reel harness, but in the event of a sudden rapid deceleration it will keep us firmly, safely and comfortably right where we are. (It's also a help to the first exposure of weightlessness. Experience proved long ago that the confining feel of the harness eases the transition for neophytes.)

Soon the airlocks—a system of thick, double hatches—are sealed, and we can listen to the flickering away of numbers in the count. Large oval windows provide a marvelous view outside the ship, and at the front end of the passenger compartment (the other passengers are engineers, scientists and two doctors) there's a row of television screens that provide a view directly forward, another directly aft, and a third for showing any part of our own ship. The windows, however, hold all our interest for the moment.

The last seconds flash away on the digital counter above the TV screens before us. We brace, expecting the sudden surge of power and the roar from the repulsors. But for the moment nothing happens! Then we *feel* it, the barely noticed sway as the ship lifts, and the slight acceleration that pushes us down gently into the thick, yielding seats. The "roar" is a deep thrumming sound overlaid with a high-pitched but still not unpleasant whine from the tremendous energy pouring into the *Evans* and being hurled away beneath us. It doesn't seem possible but the great spacecraft is rising as gently, as smoothly as a balloon floating upward. Unless we look through the oval ports it would be impossible to tell that the Earth is falling away beneath us.

The immediate view of the power grid slides away and the buildings of the sprawling spaceport complex come into view, receding swiftly and growing smaller in size. Then the mountains are level with the horizon and the sky above us starts to deepen in color. Before we know it we're at 70,000 feet and we can just begin to see the curvature of the Earth's horizon. The viewports start to darken, an automatic polarization of the "spaceglass" that protects us when we push above atmosphere, and the sun becomes an eye-stabbing atomic furnace in the sky.

PLANETFALL

A glance at the information screen shows us what our eyes and senses refuse to believe. All we hear is that steady, deep-throated hum, but we're already at 230,000 feet and thirty-five miles distant from the launching grid! Now the planet below is a great curving line, etched starkly along the horizon by a brilliant white edge of atmosphere, and absolute black beyond. We can't see any stars. They're out there, all right, but our eyes haven't had the chance to adapt from the bright sun of lift-off. So all we can see is black, and then, sliding into view on the forward screen and visible soon through the window, is a suddenly not-so-distant moon. It seems that way, at least.

Abruptly we seem to be thrown forward in our seats. We were so fascinated by the view and the flashing numbers on the information screen we forgot to pay attention to the warning of power cutoff given us through the cabin speakers. We're already eighty miles high and doing 15,000 miles an hour! It just doesn't seem possible . . . everything has been so smooth and steady. There's a brief transition period when the energy beams cut off and the nuclear drive cuts in. When the latter comes to life there's no mistake about it. Far behind us a circle of small nuclear engines is hurling a thin plasma of glowing particles from the ship, and we're accelerating again. Then, the second shutdown, and we keep our eyes glued to the front screen.

Surprisingly close to us (rendezvous is old hat by now) is an enormous structure of steel girders, spheres, cylinders, plumbing, all sorts of antenna, and flaring energy chambers—a nightmare that can't possibly be a *spaceship*! But it is precisely that, a great assembly brought up from Earth piece by piece and bolted and welded and sealed together. At times whole sections are removed and new sections attached as replacements. This is a true spaceship in every sense of the word. It will never land on Earth, or the moon, or any planet. But it can travel between these worlds, sliding into parking orbit at its destination. It is so big and so powerful it can be built up, module by module, for any major mission. The nuclear engines are fueled by a form of water to which chemicals have been added for the special needs of the nuclear drive. For deep-space missions, as many tanks as are needed can be attached to the massive structural members of the ship. Special equipment pods, or crew compartments, or scientific chambers—whatever is needed—are attached.

The *Evans* moves in slowly to the giant, heading for a cylindrical compartment with a ring of glowing lights. This is the boarding chamber where we'll transfer to the giant. The name of the huge ship is on the tip of your tongue, then you remember: *L. Gordon Cooper,* named after the astronaut who flew that ancient, tiny tin can they called Mercury 9, and commanded the Gemini 5 mission. Compared to this massive assembly, where space flight is concerned, those were the days of the Wright brothers.

The docking is completely automatic. Radar beams, computers and automatic pilot attend to the firing of small thrusters. There's a slight bump as the *Evans* soft-docks with the great ship. Magnetic clamps grip the two ships, then powerful steel probes from each ship lock onto the other, and the seal is complete. A green light flashes on—we are cleared to board the *L. Gordon Cooper*.

But not that quickly. When we release the straps we find ourselves floating away from the seats . . . we're weightless! Quickly we pull ourselves back until the velcroboots we're wearing hold us to the floor. Walking is ridiculous, disjointed and clumsy. But after a few hours under zero-g, when we're used to it, we'll be permitted to do some "directed floating," just like the space veterans who man these great craft.

We'll be in orbit for fourteen hours, enough time to stare again and again at the great planet beneath us, the continents and oceans outlined with startling clarity. Storm systems are visible, and island chains are like beads spread across a sharply curving globe. But the most fascinating and beautiful of all is sunset, which comes with a shocking, silent rush and a terminator line separating night and day as a deep red-and-gold band racing over the surface.

PLANETFALL

Within minutes we have plunged from the blinding light of the sun into Earth's shadow.

And the best part is yet to come.

When we leave for the moon.

Chapter 8

Road Map of Space

You're standing still in space. Suspended through some magic high above the planet, and when you look down through a viewport of the great spaceship the Earth slides toward and beneath you. That's what it feels like to be in orbit, two hundred miles above the entire planet. No sensation of speed. Absolutely none, and it is then, looking down on that incredibly rich and beautiful world, that you receive the first true sensing of what has happened.

You're really out here, for now you're receiving the impression of size. The astronauts learned in the early space flights not to stare at a particular point but to look far ahead of their spacecraft, to let their eyes drink in the entire glorious scene. And just like *that*, as if someone had snapped a switch, perspective changed. There came an overwhelming feeling of the ponderous mass of the globe below. Of a sudden you seemed to rush even farther away from the planet, and the wonder of what was happening poured over you. The curving edge of the world bulks out space, and the sun casts a huge sea of glimmering reflection off a storm center of high clouds. There is blue of ocean and the white mass of the polar caps and dull orange of deserts and brown of mountains. There are touches of yellow and green and wonderful mixtures of them all, and now you get the feel of it, you learn how to do more than simply look, because you fit into your own small niche of seeing. Just like that, so far removed from everything, you understand the beauty of that world and its inhabitants, and you feel closer than you ever have, and in a new and completely different way, to the planet of your birth.

It is a phrase and a thought that sticks with you. Not simply an impression, a passing fancy, but a meaning to the words you know will be yours forever, and——

The warning tone. Ten minutes to ignition. No sudden rush for a safe berth like in the "old days." It's not necessary in this huge assembly of a spaceship. In ten minutes the nuclear drive will be brought to life, but there won't be any smashing blow of flame. Wait: you'll see it on the scanning screen in your compartment. You move to a comfortable chair and secure the seat belt. That's all. Not even a shoulder harness, because acceleration will be gentle. You're pointing in the direction of flight so you'll take the acceleration in a line of chest-to-back.

The final seconds flash away on the digital counter. Far behind you there's a muffled thump. Not the nuclear drive, not yet. Small rockets have fired. The fuel is weightless and it's necessary to "seat" the fuel in the tanks so it will flow properly to the reactors. Two seconds after the gentle thump, which you felt through the structure of the ship more than actually hearing the rockets firing, the nuclear engines come alive. This time there's a roar, but it's distant, almost remote, and again the sound is transmitted through the metal structure and then vibrates into the air of the ship so that it is picked up as a sound. Within a minute or two it becomes a normal background, for the vibration is faint and the sound doesn't intrude.

Your eyes are glued to the television monitors. Outside, along the metal structure, are mounted the scanners, and they're flashing to the monitors the ghostly sight of the nuclear drive. Ghostly because there's no bright flame that you would expect from such mighty engines. The brilliant yellow flame of the old rockets, like the Saturn V at launch, came from the hydrocarbons of the fuel. The Titan II, burning a hypergolic mixture, showed a flame barely visible at close distance and, from several miles away, could be detected only by shock diamonds in the exhaust and by watching the engine chambers, which looked like two bright searchlights.

Streaming back from the array of nuclear engines is a pale violet glow spreading swiftly into a plasma that finally disappears from view. It is like a gossamer veil being unrolled behind the great spaceship, and it gives almost no indication of its energy or the effect it has upon the ship. But a glance at the digital panels by the TV monitors tells its own story. Before the reactors poured their energy to the engines, orbital speed was just under 17,400 miles an hour. Now it's 19,876 and increasing steadily. And that's how the ride will go, the big ship accelerating with a force of only thirty percent of the gravity you feel on the Earth's surface. A comfortable ride that carries the spaceship well around much of the planet, steadily increasing the distance between you and the surface so far below. Outside, as seen through the monitors, the ship races through the shadow of Earth. With shocking suddenness you're out of shadow and plunged back into dazzling sun. You don't see the sun directly, but it reflects with painful brilliance from the silvered and white surfaces of the ship.

The gentle vibration stops. No sound from the engines. But you know what to listen for. There, far behind you, the engine chambers are cooling, and, as the metal contracts, it creates a strange symphony of noises. You're coasting now, riding an invisible curving line from one world to another. The ship hasn't reached escape velocity—that is, enough speed to escape forever from the dominance of Earth's gravity. Instead, it coasts toward the moon with steadily decreasing speed, never quite outrunning the superior pull of Earth. That's the way it's been planned. Ninetenths of the distance between the two worlds the gravitational pull of the moon becomes greater than that of Earth. For a few moments you and the ship and everything within its hulk are at the neutral point where the gravitational attraction of each world is equal. It changes, of course, since the distance between Earth

PLANETFALL

and moon is never a fixed figure. The mean distance between the two worlds is 238,860 miles, but it comes as close as 216,240 miles, and reaches out as far as 252,710 miles.

The neutral point, where the moon's gravity becomes stronger than Earth, is the invisible line where the spaceship ends its deceleration. Now the ship begins to accelerate, increasing its speed until—if the engines weren't fired to slow down—it would reach the moon at about 6,000 miles an hour.

At any rate, the moment the nuclear drive shuts down, once again you are weightless. And you'll be weightless for another three days, the time it takes to coast to the moon. The trip could be made in a shorter time, but that means burning more fuel. Not too much of a concern these days with the power beams that lifted you into orbit, but fuel production on the moon is a bit behind schedule. This same ship will pick up a massive fuel load while it orbits the moon. The full requirement won't be ready for another four days, so there's little use in accelerating—and decelerating—to make the trip in just under two days, and then spending idle time in lunar orbit.

The idea of producing fuel on the moon for ships outbound to the other planets would have been a wild dream of fancy in the old days of Apollo. Every single pound had to be accounted for in those days. It took the Saturn V sixty pounds of fuel to send a single pound to the moon. When the rocket stood on its launch pad with a total weight of 6,400,000 pounds, all but 400,000 pounds was fuel.

The nuclear engines use water, mixed with chemicals, as their fuel. A long time ago—soon after the power grid for boosting into orbit was developed—engineers decided it would be cheaper (in terms of energy) and far easier to produce that water right on the moon. By vaporizing lunar rock and moving the vapor through an elaborate system of chemical separation and reclamation devices, it was possible to extract water from that rock. No water or ice in natural form was ever discovered on the moon, and so the idea of extracting water when there wasn't any raised many eyebrows. But the basic constituents were included in the rock itself. The key was to get enough energy to do the job. With the power grids a reality, it became feasible to send several nuclear reactors to the moon. These were buried far from the manned stations, and were activated, and more than enough power was made available to do the job.

That power led to the next major jump—setting up the great power grids on the moon itself. Once this was done it would be easy and inexpensive to send ships off the moon's surface and into lunar orbit. Without atmosphere to attenuate the power beams, and because the moon's gravity is only some seventeen percent of Earth's, the power beams were able to place moon-departing ships into lunar orbit or send them on their way back to Earth.

What really counted was that it didn't take any fuel to get from the moon's surface into lunar orbit. Getting back down required rocket engines and fuel, but a ship would be so much lighter after it disposed of its cargo that the drain wasn't prohibitive.

You can see the "payoff" in this system. The nuclear ships to sail to the other planets, because of their great mass and the need for enormous energy from their nuclear drive, need great quantities of water as fuel for the reactors. Lifting thousands of tons of water from the Earth's surface was a serious undertaking for just *one* spaceship. Producing that water on the moon and boosting it by power transmission from the moon's surface (where it "weighed" only one-sixth of its weight on Earth) meant fueling the planetary ships for only a fraction of the cost.

Right now, fuel production is almost complete for the flight of this giant, the *L. Gordon Cooper,* to the planet Venus. Almost— just about two days behind schedule. So the decision was made to reduce the speed and time of the flight to the moon. It will get there one day before the fuel supply is ready, which means you'll have a full day in lunar orbit before the spaceship is ready for the journey inward—toward the sun—from Earth's orbit.

PLANETFALL

Right now there are three days ahead of you, without much to do. A perfect time to learn the secrets of the road maps for space. And that calls for a meeting with the astrogator of your spaceship.

Every spacecraft ever built, from the first Vostok and Mercury "tin cans," has always had one thing in common—a master navigational display made up of many individual instruments integrated into an electronic miracle known as FDAI. The little giant with the full name of Flight Director Attitude Indicator. The early models that flew on the old spacecraft, all the way into the Apollo, had nine separate instruments to make up the FDAI. The master display aboard the *L. Gordon Cooper* has thirty-seven such instruments, all slaved to a highly advanced computer, which in turn can be ordered to change its actions by a human programmer, the astrogator.

The FDAI gives him the equivalent of eyes in front, behind, over, beneath and inside-out of everything to do with the position and attitude of the spaceship, no matter where it may be within the solar system. If the astrogator understands what the instrument presents to him there's never any question of velocity relating to the sun, to Earth, the moon, to Venus or anywhere else in the system of the sun and all its planets.

The FDAI's heart—for the astrogator—is a sphere nine inches in diameter. The scientists who designed the instrument kept in mind the needs of the man who would be studying the FDAI. They started with the concept of the old and, familiar to all pilots, artificial horizon, the gyroscopic instrument that airplane pilots use to fly under instrument conditions. Studying the artificial horizon, a pilot within clouds could always tell exactly what his airplane was doing in terms—relative to the horizon he couldn't see with his eyes—of banking its wings, diving or climbing, or any multiple of variations of those maneuvers. From this instrument, indispensable to survival and essential to navigation, the scientists went on to develop the instrument that would do the same for the astronaut, the man who flew so high that horizons became pale, and distant curves were without meaning. And that was a staggering problem to overcome.

An aircraft instrument operates on the principle that the horizon of the Earth is always present, no matter how high or fast the airplane flies. Spacecraft ascend, literally, beyond any horizon. Therefore, a substitute must be found, an artificial reference point.

Only, that artificial reference point must always be there whenever the astronaut needs it. At a single glance the FDAI must tell the astronaut whether his ship is aligned above or below, to the left or to the right, or twisted in a combination of attitudes about that reference point.

Which brings us back to the nine-inch sphere. At different moments the plastic sphere can represent the Earth or the spaceship. It can also represent, when necessary, *both* Earth and the spaceship, as well as the attitude and movement of the latter. During a flight between Earth its moon, the astronaut (or astrogator, who navigates through space) has to shake off all his prior dependence on the artificial horizon he learned to use as a pilot. Without that horizon he shuts off any reference to it and learns to think in terms of an imaginary reference.

It must be a frame of reference that is space-stable. At first such a concept is impossible. But only at first, and if the man who hopes to navigate a spaceship wishes to overcome what seems impossible, he goes all the way back to the basic elements of astronomy, advancing as his next step to navigation across the face of the Earth by using astronomical references, and then, *only* then, starting to consider the mind-squeezing reality of theoretical references when the Earth is left far behind.

The only way to navigate in space is to discard everything that's been normal to you on Earth. There's no argument about it. To navigate in space you must adopt nonhuman attitudes. Space is a never-never universe where straight lines don't exist and where four dimensions are an everyday fact of life. Theory

PLANETFALL

becomes reality, for the very good reason that reality vanishes.

For the moment let's return to the first step in astronomy, which is no more involved than standing on the Earth and looking up into the night sky at the stars. You can do this and look out upon far-flung galaxies and swirling nebulae. And somewhere out there is the key you need to unlock the problem of space navigation, of learning how to read the road maps between worlds.

You ask that nagging question: Away from your home planet, what *can* you use for the mythical but indispensable space-stable reference point? When you think of a reference point, natural instinct brings to mind landmarks, or radio navigation, or the horizon . . . but you stop all that quickly, because in space none of them mean anything.

Okay. Back to basics and looking up at the night sky. If you think about it, you realize you're seeing the heavens as a vast and hollow half-globe. Now, right here and now, if you'd like to make the grade as a spacefaring Columbus, you need to use your imagination. You start with an artificial picture of reality. Because to you—looking out into the heavens—the rim of that hollow half-globe always sits on the horizon. And you're standing in the exact center of the globe.

Once you establish this in your mind, you're ready to weave theory in and out of truth, like weaving a basket of both real and imaginary straw. You're aware that you really *aren't* the center of the celestial globe. But for all practical purposes, mainly because you're *here* and not somewhere else, you're actually creating your own artificial universe, with yourself as the center. And where space navigation is concerned, that's not ego—it's the only way to get from here to there and back home again.

As the observer of the starry sky you must chart the position and the movement of the platform on which you stand—Earth. Then, you do the same with every other object in space. This makes for some thick volumes long before you're through, and that's why navigators *and* astrogators have such thick and heavy

ROAD MAP OF SPACE

almanacs. Without them, celestial navigation (or astrogation) would be impossible. Immediate reference to star catalogs is as essential to navigation as breathing is to waking up tomorrow morning.

Now you can move along step-by-step. You've got your charts (almanacs) of *what* is moving *where,* and *when,* throughout the visible universe. This gives you the information with which to build a model along which move particular stars and the planets. Remember, you're imagining the paths along which move the celestial bodies of the sky. And you've got to have a schedule of *when* those bodies move. Finally, and above all else, you've got to keep in mind that everything is always shifting in relationship to everything else.

(You're beginning to have some idea of why computers with their vast memory banks are so vital to making sense out of the road maps for space . . .)

You must accept that all space activity is in four dimensions. Every object in the universe is always moving in relationship to every other object, so you can't consider anything at all without *time.* You can get away with ignoring time on Earth because if you stop moving, so does everything else. Mountains, cities, oceans, remain fixed in their reference to you, no matter where you are.

But not in space. And this is absolutely critical to understand. When you're effectively brainwashed so that you'll never forget this difference in navigation in space—this element of four dimensions—you're ready to start figuring out the routes you can travel to move from one object in space to another.

Let's not get over our heads with Venus. You can stick with the flight between Earth and moon. And you start with a brain-twister.

You need to erect a perpendicular line right in the center of the plane created by the orbit in which the Earth travels around the sun in one year.

A perpendicular line in the center of the orbital plane.

PLANETFALL

Time for a deep breath. Maybe the terms are just throwing you, that's all. Make a mental picture of the orbit of the Earth around the sun. Easy enough for that big nearly-circular swing. Next, think of that orbital line as a dish.

It's not a flat dish, like a pane of glass. If you could see the line drawn by the Earth's orbit, you'd also see that, using the sun as your center, the plane of that orbit curves upward slightly. That's from the sun outward to the orbit itself.

Well, you don't *have* to use the dish concept. That upward curving surface could throw anybody. You can ease off the mental pressure if you think of the plane of the orbit as a great, flat circle.

Okay. As soon as you've figured out the plane—the flat dish created by the Earth's orbit around the sun—you discover that everything isn't balanced as neatly as it should be. To go anywhere in space from the Earth, obviously you must use the Earth as your launching platform. Careful—it's lopsided.

Earlier, you drew a perpendicular line directly in the center of the plane created by Earth's orbit about the sun. A perpendicular line is true vertical. And the axis of the Earth, of course, should match this perpendicular line. It should be straight up and down compared to the orbital plane of the Earth. It *should* be, but it *isn't*.

Picture the orbital plane of Earth as a tremendous sheet of glass. It gives you a handy and definite mental picture. Now when you look at the axis of the Earth, through its poles, you can see how badly it tilts. In fact, it leans over about twenty-three and a half degrees from the vertical line you went to so much trouble to establish.

So, *now* you start building navigational castles in your head . . .

You've just learned that things are a bit cockeyed. The way out of this dilemma is to alter both your actual and your theoretical models of the heavens. You must consider the axis of the Earth not only in relationship to the orbital plane—that sheet of glass—but also in relationship to the rest of the celestial family. You may be so unhappy about all this that you decide to shuffle things around until your brain puts everything in its place. After all, if reality can be a bit cockeyed, why can't your model of reality be the same? When you're all through you've done it. You have a celestial axis you can use as a definite reference point.

Well, we'll admit that it has its limitations, because it tilts. Following the example set by the Earth, your celestial axis deviates twenty-three degrees or so from your perpendicular line. At this point you may want to kick everything into a corner and go home, because after going to all this trouble to set up your perpendicular line, nothing you do from this moment on will match that line. But hang onto it. It's important.

You now have a model of the universe and a celestial axis, but everything seems to be askew. It's all leaning away from the vertical, and the whole universe seems to be cockeyed. But you've made progress, and now you can make a vital decision.

You decide that the real universe *is* tilted. You've quit fighting and you've accepted the fact that you must change your way of thinking—where it relates to concepts of life *on* Earth.

If you really want to make flights away from Earth it doesn't matter at all if everything *looks* wrong. It doesn't matter that the whole universe is leaning drunkenly against the lamp post of the perpendicular line you drew. What does matter is that you're accomplishing things, taking one step after the other and, believe it or not, some order is starting to emerge from the models (concepts) you've built up, torn down, and built again. You know *why* you've had to do this.

Which means you now have some definite reference points. You're no longer groping in the dark. Now, at long last, you can *measure the movement of objects against objects.* You're able to hang onto your fixed reference point. So it's tilted. So what? You're about to move into the spaghetti of circular and elliptical movement, and that old reference point, tilted or not, is going to become a very friendly neighborhood.

PLANETFALL

Now it's time for a small miracle. What you've done so far, despite the headaches that went with it, was to establish your reference points. Until this moment everything has been confused. But with those reference points as a solid anchor, you look up to discover that everything has dropped perfectly into place. Instead of chaos, the universe shows its extraordinary sense of order. You blink as though until now you've been blind and the sun has just started to shine.

Most problems in science or life can be solved once you've got solid footing beneath your feet. That's what you have right now. You've got certain fixed references with which to guide yourself. If you look away they'll still be there when you turn back. *Now* you take another look from the Earth toward the sun. It's possible to see many things at one time—*but they all fit into place.*

First, you see the sun, just as you did before, but now you see the sun *with the stars as a background.* Now's the time for that small miracle we mentioned. Because you have solid footing beneath you and you know how to look, you discover that the *apparent path* of the sun among the stars leads through a route that it follows over and over. There's your key. Something is happening that repeats itself again and again. Repetition means reliability.

So you know you've stumbled onto something. Actually, you have just learned the basis for accurate guidance between the Earth and the moon.

For the first time you have established what navigators (and astrogators) call the space-stable frame of reference. In other words, that frame of reference is always out there and it always happens the same way. If you know where to stand, and what to look for, you can pull off one of the best tricks of all time: you change what's theoretical to what is real.

You need some name for what you've discovered. The apparent path of the sun among the stars? You call this the *ecliptic.* What brings more confidence to you now is that in the course of one year the ecliptic will always be a familiar avenue of travel for the sun. That old great feeling of reliability. And then you notice something else. The movement of the sun along the ecliptic *always leads through twelve constellations.*

At this point you're transferring what you see to your road map—you're beginning to chart the celestial highway. And you find out that while the problem of traveling from the Earth to the moon and back again may be a relatively new matter, that map of the celestial highway really has been around for a long time.

Right—it's the Zodiac. A group of twelve constellations. The same twelve constellations that the sun always traverses—moves through—as it slides along the ecliptic.

Aries, Taurus, Gemini, Cancer, Leo, Virgo, Libra, Scorpio, Sagittarius, Capricorn, Aquarius and Pisces.

Okay. Time to take stock of things. You've got your reference points. You know about the sun moving along the ecliptic. You know how the Zodiac fits in. You're ready to slip another piece of the puzzle into its proper place.

First, keep in mind something you learned before. The Celestial Axis deviates by a little more than twenty-three degrees from the perpendicular, that line you created originally for a reference.

Now you try to reason out more details. If the Celestial Axis deviates by twenty-three degrees or so, then the Celestial *Equator* must do the same thing. It also tilts at an angle of some twenty-three and a half degrees.

Take it slow now. You know the Celestial Equator tilts. By itself that means nothing.

It tilts—but to *what?* What can you use for a frame of reference, a comparison?

Watch carefully as everything starts to fit. You can see that the Celestial Equator tilts at an angle of twenty-three and a half degrees *with respect to the ecliptic.* All of a sudden there's a really important relationship between the different conclusions you've drawn.

What you now have is a solid reference point you can take all

PLANETFALL

the way to the moon and back home again. A permanent road sign in space.

Of course, there's every chance that at this point your patience is wearing thin. "How," you demand, "can you use what you call a permanent road sign in space when it's theoretical? It's something that's been made up, a mental image. How do you use *that* as a signpost? After all, can you imagine people trying to find their way around using theoretical road signs?"

The answer to your little outburst would be that you missed the point of a few moments ago. Certainly, your road sign in space is a theoretical reference point, but at the same time it's real. Because as a navigator you can always use it as a reference against celestial objects and their known positions at any one time.

Aha! But how do you know their known positions at any . . .

Right. You just remembered the almanac. It's like a flight schedule for the planets and the stars. It tells exactly what will be where at a specific point in time.

And it gives you, always, a constant reference to the timetable of objects in space. So your theoretical reference point becomes real.

Let's have that again?

Your theoretical reference point becomes real because you compare it to *what is real, both in terms of location and time.*

No matter where you are between the Earth and the moon, so long as you know where and when to look—call it where-when if you want to nit-pick—you can figure out where you are, how fast you're going, and along what orbit, and when you'll arrive at your destination. Of course, this is simplifying things a bit. The mental gymnastics could scramble any brain. That's why spaceships are built with highly advanced computers and all sorts of instruments the computers use for navigation along the celestial highway. Plus some help from the astrogator, who works with the computer, the almanac, sextants, scanning telescopes and other devices.

ROAD MAP OF SPACE

Any space Columbus who's skilled in the foregoing can also find solace in two other explanatory references—which he then learns by heart. You can consider them among the most basic facts of space operations. First, that the ecliptic is more than the apparent annual path of the sun among the stars. It is also the intersection of the plane of the Earth's orbit with the celestial sphere. And that works out as a great circle of the celestial sphere inclined at an angle of about 23 degrees 27 minutes to the celestial equator.

Any navigator or pilot learns early in his career that it isn't enough just to understand the finer points of celestial navigation. This is a science that bases its existence on more than simple mathematics. It demands an approach all its own.

The man who knows how to move along this avenue is certain to find the real secret to understanding. It's known as harmony. One day you stop poking and prodding and stumbling about through the thick forests of equations. Your head hurts, so you take a deep breath and, in your mind, you seem to step back for a different view of things. Everything you've learned up to this point has—whether or not you're aware of it—changed your perspective. You've got levels of understanding that may not even be apparent to you.

So you step back and relax and look up and let it all sink in. Not just the numbers but the beauty of it all. And that's when— it's happened just this way to many astronauts—everything locks into place.

One moment the exquisitely complex structure is fracturing your mental vision. And then, like a great orchestra that completes its individual tuning up and joins together, harmony appears. It's as if you can project yourself to any coordinate within the four-dimensional structure of space. It's a mental balancing act that every pilot and every astronaut comes to master, and the effort becomes instinct.

It's like inner seeing. You know, and you comprehend, that the movement of any one object in space affects, and is affected, by the whole. You *know* this to be so. Almost at will, now that everything has its place, you can spin in your mind the gossamer paths that flow and shift with precision from one place to another.

If you're able to understand the nature and the role of the ecliptic, you have gained the indispensable building block to the mastery of celestial navigation in space—what is known as astrogation. If you comprehend the ecliptic, you're able to enlarge upon your basic capabilities. You're ready, at this point, to move on to the second "explanatory reference," the Ecosphere.

Using your astronomical charts at this point allows you to use the Ecosphere as a navigational hinge from which limitless doors can open. The distinction between Ecliptic and Ecosphere?

They are *almost* the same. The slight difference opens those doors to navigation through space. The Ecosphere is the great circle on the celestial sphere that describes the apparent path of the sun in the course of one year. But isn't this the *same* as the ecliptic? *Almost* . . .

The *plane of the ecliptic*—ah, there's the crux of it all. For this is the plane in which the center of mass of the two-body system made up of the Earth and the moon revolves about the sun.

Reality decrees that neither the Earth nor the moon can be used as if the other were not there. Reality demands exquisite accuracy, and so it's necessary to judge the Earth and the moon as they really are—two bodies revolving about their common center of mass.

Here is another very real case where theory replaces reality. How can the Earth revolve about a center of mass several hundred miles beneath its surface? Well, strange as it may seem, the center of mass of the two-body system of Earth and moon *does* lie several hundred miles beneath the surface of the Earth.

Okay, you say. The same question again. How can the Earth revolve about a center of mass several hundred miles beneath its own surface? It's a valid question, as the Earth, of course,

ROAD MAP OF SPACE

can't revolve about itself. But in terms of navigation through space, *the center of mass is everything.*

It is the mathematical universe, and not what is seen by a man's eyes, that holds sway and is absolutely dominant.

Back to a statement we've repeated several times—but now with new meaning. The plane of the ecliptic is inclined to the plane of the equator at an angle of 23 degrees 27 minutes. What does that statement mean? Nothing, really. *Unless* you're aware of the wonder and predictability of celestial order. Quite suddenly, then, two reference points are at hand, both are specific, and both hold meaning to one another.

Which brings us back to the FDAI. Only now, that marvelous device gains a far greater clarity of focus and meaning, because it is part of a greater whole.

A two-body system, a specific center of mass, the plane of the ecliptic, *and* the FDAI—mean for our spaceship an accurate, predictable and reliable two-way highway between Earth and moon. Built upon knowledge and a comprehension of celestial fabric, cemented by skill in its use, the nine-inch sphere of the FDAI provides to the navigator of a space-sailing vessel a symbolic representation of the spaceship's orientation, with respect to a space-stable frame of reference.

The plane of the ecliptic.

That's the name of the highway.

What, you ask, happens if something goes wrong with the computer? After all, it's happened before. Computers can blow their electronic marbles. And without the computer to keep track of the mind-boggling numbers and figures of the celestial almanac, what could we do? We wouldn't be stranded in space because we'd still be in the Earth-moon system, and there's rather obvious visual reference, and there are tracking stations and computers both on the Earth and the moon to help out. But what happens far *beyond* the moon?

PLANETFALL

Everything is locked up within that marvelous package we call the FDAI.

When you look at that nine-inch sphere you're actually seeing a microcosm of the greater reality in every direction. You're seeing a perfect balance between the innards of the FDAI and the wheeling of the stars and planets in their orbits. It's all a part of the vast celestial orchestration, everything moving to the same beat.

Remember, the FDAI provides an *all-attitude* display. The small globe and its transparent overlay with index marks and needles do more than tell an astrogator a spaceship's body and navigation axes at any given moment. The FDAI also sits in judgment as a monitor of the extraordinarily complex guidance and navigation system, *and* the automatic stabilization and control system, of the huge nuclear-drive spaceship. Not only does it do all this, but it registers constantly any shift in the mass of the spaceship and, when the time comes for firing up the nuclear drive, considers any and all changes that have taken place in the velocity, mass, center of mass and attitude along the path of flight.

All the elements of the intricate stabilization and control systems are linked to the gleaming plastic sphere of the FDAI as if it were some *living* nervous system with everything terminating in this single, swollen eyeball. Whatever the *L. Gordon Cooper* does, has done, or may do—its attitude about its three axes of pitch, roll and yaw; its rate of movement about these axes; and any deviation between the required and the actual attitude—is always displayed simultaneously by the FDAI.

Yet by no means is this the full extent of the internal circuitry of *L. Gordon Cooper*'s electronic brain matter and sensors. When the scientists created the FDAI they knew they would always be dealing with some feedback of spaceship motions that might interfere with the sensing and read-out displays of the FDAI. There are always some undesirable oscillations of a spacecraft. These can come from urine and garbage dump, the effects of the solar

ROAD MAP OF SPACE

wind (especially closer to the sun than is the Earth), the firing of small thrusters for attitude control, the consumption of fuel, the changing mass of the ship—and perhaps a thousand other minute factors. What all this means is that there is a critical, continuing need to couple the working elements of the guidance and navigation system with the firing of small thrusters and any number of powerful nuclear engines for velocity changes, for attitude holds and changes.

Whatever happens with *L. Gordon Cooper* or any spaceship venturing away from Earth is always displayed immediately on the face of the FDAI.

Then—and this is the final question, of course—what happens when a spaceship is far from Earth and the computer goes out? Wait—as well as radio communication with Earth or the moon? At that point the electronic umbilical cord is severed and the spaceship is utterly, impossibly separated from any help. What happens next?

At that moment the astrogator starts earning his keep in a way he hoped would never happen. This one man, and others of the crew working under his direction, must now function as human computers. Their very presence on board the spaceship makes them available as human back-ups for attitude control, guidance and navigation, and velocity control. The main control room FDAI, should all power be lost, immediately resorts to a spare system of integrating gyros. And just as before, the FDAI will present, through its plastic sphere and display face, the crucial attitude data and rates by which the astronauts would assume direct manual control of their huge spacecraft.

That's why every ship that moves out beyond the moon has a main control FDAI, another in an emergency control center, and a third within the bowels of the ship, surrounded by thick steel shielding.

Because if you lose everything else, just so long as you understand the rules of the celestial highway system, and your FDAI works, you've always got your return ticket home.

CHAPTER 9

Time Machine

For all that has been learned about it, for all the poking and study and carving and gouging of samples, for all that has been done to strip away its mysteries and to unlock its secrets, it remains fascinating, beautiful in its own way. A world that can kill swiftly, but also a world that is willing to accept its alien life forms in their pressure suits and strange habitations that come from the planet Earth a quarter of a million miles away. And there is always the knowledge that this world, so lifeless before the first visit of men in the summer of 1969, remains a time machine, with the history of itself, of Earth, of the solar system, perhaps of the galaxy itself, locked within its rocks, hidden within its tortured surface, concealed deep beneath its crust. Everything about the moon and everything that takes place within the moon, as we learn to unravel these secrets, carries us far back in time, to that incredible beginning when the sun was just forming and the planets themselves were no more than glowing whorls of dust revolving about a newborn star.

The moon is the lodestone of planetary evolution in this system of ours, and as you drift toward the battered and beaten globe, you are grateful, as it swells larger and larger in your viewport, that you are still awed and deeply impressed and even a bit overwhelmed. And you marvel at how little we knew, and at how wrong we were, about so many things before Neil Armstrong first set his booted foot onto the surface of the Sea of

Tranquillity and said in words that will live as long as our race will endure:

That's one small step for a man, one giant leap for mankind.

Now we'll reverse our movement in time and go back to the years just before the first Apollo spaceships embarked on their journeys to the moon. Let's see just what was believed about the moon, and then what was learned, and how it changed drastically our knowledge of the most fundamental facts of our existence.

First, there were questions upon questions, all of which stemmed from certain basic knowledge about the moon. There was a kicker in this "knowledge." Much of it was as much assumption as it was confirmed fact. We had gross figures that were in urgent need of refinement, and until we could get to the moon and shave off those rough edges, almost everything we knew about the moon, or tried to deduce from our studies of the moon, had to be held in serious question.

Which meant we had a basket bulging with theories instead of hard fact.

The "rough edges" told us that the Earth revolved about the sun with a mean orbital speed of 18.5 miles a second, or 66,000 miles an hour. As it raced around its parent star, the Earth dragged the moon along with it. Our picture was clear enough—Earth moving in its stately planetary circuits while the moon tags along faithfully, circling the Earth in turn—about once every twelve times or so that the Earth falls around the sun. It's accurate enough but it still has those rough edges. If we're referring to the Earth and the moon and circuits about the sun in everyday terms, it meets our needs. But if we want accuracy in terms of the astronomer, the picture is a crude sketch rather than a detailed portrait.

When we beat our knuckles against our heads trying to fathom the intricacies of space navigation, we mentioned that in mathematical terms the moon doesn't revolve about Earth. In terms of mass, the two bodies form the Earth-moon system, and they orbit about a common center of mass that lies about a thousand miles beneath the surface of the Earth, or approximately 2,900 miles outward from the center of the planet. This may seem to be splitting hairs, but when astronomers calculated the movement of the Earth and the moon about the sun, it was (and still is) this common mass to which they refer.

Now, no astronomer before Apollo believed he knew *accurately* the mass of the moon. In those "rough edge" figures we knew the moon had a mass only $1/81.3$ that of Earth But trying to determine the lunar mass from the surface of Earth, based primarily on observations of the motions of asteroids and of the motions of the Earth's polar axis, left an error estimated to be about 0.3 percent.

Doesn't seem like much, does it? To scientists this question of 0.3 percent is a "great uncertainty," and was one of the most important "rough edges" to be refined. It was refined—by recording every minute detail of the flight of spacecraft to the moon, by determining to extreme accuracy (through laser beam reflections) the distance from Earth to moon, and other experiments. One immediate result of knowing to much greater accuracy the mass of the moon was a major improvement in the guidance and navigation systems of the robots we were sending to different planets.

Back to some basics . . . The visual scene of the moon sliding about our world, as we see it from Earth, is a movement from west to east, with a rate of movement approximately thirteen degrees in each period of twenty-four hours.

There's always been a neat coincidence about the moon. Its diameter—2,163 miles—is astonishingly close to its average speed in orbit, which is 2,187 miles an hour.

As it revolves about Earth the moon also rotates. The rate of rotation is one complete turn on the lunar axis in every $27\text{-}1/3$ Earth days. Rotation and revolution come close to matching one another, so we have the situation familiar to us all—the moon presents the same face to us at all times, and, until the first probes dipped around to the lunar far side, we could only guess about conditions we were never able to see (and, again, we were to be surprised).

We said that rotation and revolution almost match, but not quite. The moon also has a slight wobble on its axis, and as it moves through this slightly drunken spin, it exposes a bit more than half the lunar surface. Just enough to whet our imagination. Fifty-nine percent isn't enough.

Each lunar day and night—again, such facts were the results of observations from the first days of astronomy—lasted not quite fourteen Earth days and nights.

Whatever we knew about the moon, there was so much more we didn't know. For example: What was the surface composition of the moon? How was the moon formed? Does the moon have a magnetic field? Is there a molten core to the moon? Are there active volcanoes? Are there moonquakes? What about water or ice deep within the moon? Does gas seep to the surface? Are there any life forms on or within the moon? Did the moon ever have active volcanoes? What formed the craters? What material made up the huge basins, or "seas," on the moon? What does the far side look like? What are the lunar rays? What created them? Are there rocks and boulders strewn across the surface? Is there color on the surface or is it all gray?

Just the one subject of surface material was enough to have scientists pounding their fists against meeting tables. Men of great repute stated, with absolute conviction, and in violent opposition to their colleagues, that the surface of the moon was of hard rock: jagged rock, lava formations, deep and sandy depressions, thin dust, deep dust, *no* dust, crunchy as peanut brittle—ad infinitum.

One of the best examples was provided in 1958 by Professor Thomas Gold, of the astronomy department of Harvard University, who painted a vivid word picture of conditions on the lunar surface. According to Dr. Gold, the surface of the moon must have vast areas of deep and dangerous dust. Dr. Gold said that "it would appear that the depth of this dust should be in many areas of the order of one kilometer . . . it is appropriate to discuss in what condition this dust surface might be . . ."

According to Dr. Gold, then, we could expect to find a sea of dust *three thousand feet deep* in many areas of the moon.

Dr. Fred L. Whipple, director of the Smithsonian Astrophysical Laboratory and professor of astronomy at Harvard, was one of the few scientists who came up with a fine, accurate description of major lunar features when he wrote that the "slopes of gross lunar features, except for occasional faults, are always relatively mild and are extremely gentle in the maria . . ."

Dr. Whipple was in strenuous disagreement with Dr. Gold's conclusion about the actual surface materials. Dr. Whipple predicted the surface materials would be "perhaps comparable to sand," and he went on to state that "loose dust on the lunar surface is practically nonexistent. . . . It should be sufficiently cemented or consolidated that it will not blow out dangerously in a rocket jet. To the human (encased) foot or under vehicles the surface should be 'crunchy' and allow minimal imprint."

Then we had V. A. Firsoff's studies of lunar physics, which led him to conclude that both Gold and Whipple, in their essentials, were far off from the reality that men would find on the lunar surface. After a lifetime of astronomical observations of the moon, Firsoff said flatly that the lunar surface was characterized by ". . . rough ground, largely impassable to any landborne transport, even at first sight, and it may well prove still more impassable than it looks."

Firsoff and his adherents insisted that the lunar surface must be pumicelike in nature, with jagged rocks and edges of rock formations, and the surface, underfoot, of a density less than water.

So what we had, before the first robots such as Luna and Surveyor landed on the moon, was an extraordinary variation in scientific predictions by the leading men in their professions in the world. It extended from a surface dust thickness calculated at one-fifth of an inch to 3,000 feet! By and large, however, the majority of scientists felt there was a thin dust layer on the moon. They also believed that constant solar radiations had cemented the dust into the "crunchy" substance forecast by Dr. Whipple.

In the early days of probing the moon with robots, every precaution was taken to assure that the payloads that were to impact the lunar surface were sterilized so thoroughly that no terrestrial organisms could hope to survive this treatment. The first Pioneer payloads, atop the rocket standing on its launch pad, were even treated by scientists and engineers in surgical gowns, caps, gloves and masks. The idea was that terrestrial organisms could contaminate the lunar surface. Not in terms of spreading dangerously, but to an extent that might confuse or mislead initial research on the moon.

Dr. Harold C. Urey, of the Institute of Technology and Engineering, University of California, reacted in scathing fashion. "I should like to express it as my personal opinion," he said, "that the enormous amount of talk in the newspapers lately about contaminating the moon with biological materials is mostly utter nonsense. Conditions of the moon are such that no biological materials could multiply there." He said that even if there were an accident that spread biological organisms, "within one month all bacteria would be dead . . . they would dry up completely."

Dr. Urey also stated that the difficulties of navigation to the moon were so enormous that the first landings would be a proposition of landing just where a manned ship might be able to descend safely—anywhere. He said: "The chances of finding again the same spot on which the first landing was made are rather small. It would be necessary to spend some years exploring the moon before it could be found."

PLANETFALL

Apollo XII was going to give Dr. Urey some unpalatable food for thought . . .

Let none of this give the impression that these were conclusions reached along any hasty route, or that they represented anything less than serious, scientific findings based on the best knowledge available to the individuals involved. It's just that not too many people understand that what we are given as scientific fact is, most of the time, nothing more than opinion or conclusion, often reached despite a distressing lack of solid information from which to shape that conclusion.

It's important to keep this in mind, because it enables us to also keep that "grain of salt" in our desire or willingness to believe what eminent men of science tell us. It wasn't so many years ago that some of the more respected scientists in this country ridiculed, and sometimes attacked violently, those men who wrote of rocket flight beyond this planet. It was so obvious, so utterly clear, so beyond all question, granted the scientists, that a rocket could never work in space, that whoever advanced such a proposal was a contemptible ignoramus. And why wouldn't a rocket work in space? Because, the scientists explained in icy undertones, there was no air in space against which the rocket could push.

Today such a statement would bring on shocked silence that anyone could make *that* stupid a remark! We know rockets function in space through Newtonian laws of action and reaction. But we—the man in the street as well as the scientist—know this because other men have sent rockets away from the Earth. Therefore, it's *obvious*.

Sure. *Now* it's obvious.

The problem about placing boundaries in the future is that whenever we insist something can't be done, we're operating under serious restrictions. *We're basing our decision on what we happen to know about the subject at this moment.* And that can be an appalling error, because if you are aware of only a tiny slice of the picture, you can hardly expect to be an authority on the entire subject. Yet some men, with no more than this tiny slice, take it upon themselves to act as authorities. *Most of the time they don't know they're seeing just the smallest fragment of the subject.*

Man's long struggle to fly within his atmosphere provides some of the clearest examples of this problem. Scientists insisted a heavier-than-air machine could never fly because of what it was—heavier than air. A balloon could fly because the gases contained within its envelope were lighter than the surrounding air and therefore provided a natural buoyancy to the balloon and its basket. But this was clearly impossible for an airplane, a flying machine, to do. Since it was heavier than air, and it did not use the buoyancy of lighter-than-air gasses to fly, it could never stay above the ground.

It's so *obvious*!

Well, those same scientists did not understand the nature or the characteristics of air itself. They didn't understand that air is a fluid medium. They had no idea that air under movement can be made to create pressure zones. If they had known the characteristics of the weather engine that is our atmosphere, then they would have known that a high-pressure zone always tends to move into, or to displace, a zone of lower pressure.

Okay. So we take an airfoil—our wing—and we move it forward through the air. You can do this by throwing the wing (in its assembly as a glider) off a cliff, so that gravity gives it acceleration. Or you can use a rocket to hurl it forward. Or the speed of a horse dragging a rope to which the glider is attached. Or a propeller that's powered by a piston engine, steam, or a rubber band. It doesn't matter. Just get the airfoil moving forward through that fluid medium we know as air.

A wonderful thing happens. The flow of air past the wing is shaped by the curve of the wing. The result of all this curving and shaping is that as the wing moves through the air there is normal pressure beneath the wing and a zone of lower pressure above the wing. What's that fundamental law of weather? A high pressure zone will always displace a zone of lower pressure. Right on. So the air beneath the wing keeps trying to get into that zone of lower pressure above the wing. It tries, but it can't get there because the wing is in the way. What happens? There's constant pressure, pushing upward, against the entire bottom of the wing. If the total force of that pressure—we'll say it's three thousand pounds—is being applied against an airplane that weighs two thousand pounds—we get?

Lift.

And the airplane flies.

Simple, once you know how. But if you don't, and you're still unaware of such things around you, you can make the definite statement that a heavier-than-air machine *cannot* fly. And until someone comes along and breaks the rules, you're right.

Even if you're wrong.

In the first chapter of this book I pointed out that far and away the majority of those who studied or wrote about the surface conditions of the moon in pre-Apollo days (including myself) had fallen victim to the spell of jagged shadows stretching with such foreboding across the lunar surface, and had mistakenly assumed that because the shadows were jagged, so were the mountains and crater rims that made up their source.

One of the more delightful aspects of studying strange and distant worlds is that you never know where accuracy will stray into the picture. And when it does, it's not often recognized, except after a long period of time and much effort. Such is the case with the lunar surface. Those of us who reveled in stark descriptions of this dangerous world (aided in no little way by such men as Firsoff, who described the moon as savage in its features) would have done well to have read the reports of the Aeronautical Chart and Information Center of the United States Air Force. The ACIC since 1958 had been engaged in an exhaustive program to produce accurate charts of the visible portion of the moon, and was using photographs and other materials from observatories throughout the world, implementing these pictures with elaborate chart-making equipment and procedures. The ACIC, with contracts to astronomers and observatories, divided the moon into forty-four separate areas to receive intensive study. Each one of those areas had to be photographed at least four times, and each time under different lighting conditions.

At the conclusion of this task the ACIC had prepared a picture of the moon in what could be described as being in "violent conflict" with the findings of most astronomers. Before we get to that definite description of the lunar surface, there's a fascinating report that describes, in part, how the job was done and the enormous problems encountered in such a task. The material quoted below will provide the reader with a rare look inside such a project and also help to emphasize the problems in space navigation we investigated in the last chapter. The material immediately following is from an Air Force study of the ACIC Lunar Charting Program:

> An instrument called the Variable Perspective Projector helped "crack another nut" in the project. Considerable distortion results from photographing a spherical object on flat film. These distorted images require rectification to bring them into proper relationship. The moon job was done by projecting negative pictures of the moon into a curved mirror, thus converting the light rays, which normally narrow down to a pinpoint, into parallel light rays. The image was reflected on a sphere and rephotographed.
>
> Another problem was posed by the phenomenon called libration, or the oscillatory motion of the moon as it courses through its orbit. Because of these oscillations—both real and imaginary —eighteen percent of the lunar surface is alternately visible and invisible to earth observatories, while forty-one percent remains always visible and the other forty-one percent is never visible.
>
> Because the moon's axis is tilted from the plane of its orbit, 6.5 degrees beyond one pole is visible for about two weeks, then the same extent beyond the other pole in the next two

weeks. This is libration of latitude, or optically an up-and-down effect.

Then there is a side-to-side or longitudinal libration. This is due to the fact that, while the moon, like our planet, rotates on its axis at a regular rate, the velocity with which it revolves about the earth is not uniform. Its path is an ellipse and its speed is greater than at perigee, the closest approach, than at apogee when it is most distant. At perigee, observers can see a little more of the moon around one side and at apogee, a little more beyond the other side of the always visible lunar disc.

A third variety is the diurnal (or daily) libration which enables an observer to look slightly "over the top" of the moon when it rises and later a bit "under the bottom" as our satellite is setting. Actually, this becomes possible because of the libration of the observer's earth platform and not of the moon.

Finally, there is the libration caused by what is known as precession—a wobbling of the moon due to a gradual shift of the tilt of its axis with respect to checkpoint stars. This may be compared to the motion of a top which is not spinning perfectly upright but "leans" from the perpendicular, or a gyroscope similarly tipped. If the axis of rotation is related to something fixed, it traces a circle.

The moon is also subject to an interplay of gravitational forces originating from solar system components other than the earth. These cause an alternate acceleration and braking of the orbital motion our satellite would normally have without such interference. The resultant irregularities are known as perturbations. Some produce an extremely slow and steady change in the orbital pattern of the affected celestial body, others cause relatively short intervals of oscillation. Lunar cartographers can find no joy in that the moon represents one of the toughest perturbation problems known. Tables have been worked out for predicting most of the irregularities it is subject to and these tables fill three quarto (9.5 x 12 inch) volumes totaling more than 360 pages.

In the whole moon-mapping project, ACIC's main objective was to get the most out of available lunar information. The production of these charts meets the first challenge in photography. As our nation's space vehicle operations require more and more

TIME MACHINE

detailed charts and aerospace navigation information, the challenge will increase. Coping with challenges has provided ACIC with the experience that made this moon-mapping project possible. More challenges will lead to more experience, consequently, wherever man plans to go, ACIC chart makers are confident that they can, cartographically, precede him.

And that picture of the moon forecast by ACIC so many years before Apollo? In the exact words:

> "After many geometric calculations, cartographers at the Center have given us a new idea about the topography of the moon. Contrary to our concept of craggy cliffs and abrupt craters, most of the moon's complexion is relatively smooth, with gently sloping hills and ridges."

We should have paid closer attention . . .

Now for the "new" moon, or what might properly be called the post-Apollo moon.

It wasn't a cold and dead planet. It didn't roll through space as an unmoving, unchanging sphere. Absent from its surface was atmosphere and biological life, but what was missing across the rolling terrain was found within the globe. Moonquakes shook the small planet; they shook it frequently and with regular cycles and with a type of deep tremor never known on Earth. And there was heat—there *is* heat—within the moon, detected and verified by instruments placed and left on the surface. Toward the core of the moon there is a sudden rise in heat that likely reaches to 800 degrees F. It may be volcanic or radioactive in nature or simply the result of awesome pressures; that still must be determined. But heat and quakes there are.

Spreading far beneath the lifeless surface we see, and upon which men have walked and ridden, is an enigmatic material, best described as honeycombed or of packed rubble. There are vast sheets of dense material sliding across and over and beneath one another—the mascons, the areas of dense matter that tweak and dip the orbit of ships passing overhead.

It began on December 21, 1968, with the 7:51 A.M. launch of Apollo VIII. Borman, Lovell and Anders rode the first Saturn V ever to carry men away from Earth, with a speed of 24,200 miles an hour and headed on a curving highway that ended with a beautiful orbit about the moon of sixty miles. On man's first excursion away from his home planet he spent Christmas in orbit about the moon, and three men read from the Book of Genesis as their gift to an awed world a quarter of a million miles away.

Apollo VIII spent twenty hours and ten orbits about the moon, and from that moment on everything was changed. The side of the moon facing Earth is covered with huge mares, or basins, of relatively smooth and dark material. They were absent on the lunar far side, which no human eye had ever before seen. The "dark side" of the moon was a hellish chaos, showing signs of terrible violence, of massive explosions and sledge-hammer blows. In every direction, that violence had been frozen in some cataclysmic past, and now we were seeing at first hand what no one had been able to predict. That alone told us much, for those first looks, captured in memory, on film and on television, made it clear that the lunar crust on Earth side was much thinner than on the far side. After the ten orbits, which also proved the existence of great masses of debris and boulders (some scientists said that such objects could not exist on the moon) from violent impacts with objects from space, Apollo VIII fired up its main engine while on the far side of the moon. Would that engine work? If it failed to burn as needed, the Apollo VIII crew would be marooned to die slowly in lunar orbit. Then a radio message flashed to Earth: "Please be informed there is a Santa Claus." Apollo VIII was on her way home.

After their return, before a joint session of the Congress, Frank Borman spoke quietly in words that will last as long as men remember the first voyage to another world: "We stood on the shoulders of giants. Because how can anyone think of Apollo

VIII without thinking of Galileo, Copernicus, Kepler, Tsiolkovsky, Oberth, Goddard, Kennedy, Grissom, White, Chaffee or Komarov. If Apollo VIII was a triumph, it was a triumph of all mankind and we acknowledge it as being such."

Apollo VIII proved some other points absolutely critical to future flights. Man could withstand the awesome speed of 25,000 miles an hour as the spacecraft ripped back into the atmosphere. For the first time man had been under the dominance of the gravity of a body other than Earth, and he had proved the magnificent accuracy and control of his navigation and guidance systems between worlds. For the first time, men raced through the intense radiation belts of the planet and went beyond the protective sheath of Earth's magnetic field. The future boded well.

On July 16, 1969, at 9:32 A.M., Apollo XI, with the landing ship *Eagle,* crashed away from Earth, built up speed to 24,000 miles an hour, and was on its way for the epochal landing on another world. On the afternoon of July 20, Eagle reached the lunar surface. Neil Armstrong, in his usual quiet voice, sent a message to a breathless world:

"Houston, Tranquillity Base here. Eagle has landed."

The age-old question was about to be resolved. The lunar surface in the mare basin, known as the Sea of Tranquillity, was a flat expanse only in that it lacked major topographical features. Nowhere was it—despite photographs from Earth, from robot probes, and from Apollo VIII and X—smooth in the sense that it had an uncluttered or unlittered surface.

There were craters everywhere, including one almost as large as a football field, into which Apollo XI was heading, until Armstrong took over manual control and steered to a safer point of descent. They touched down with no physical sensation, but there was no question that even the "gentlest" area of the moon was pitted and gouged with craters. The prediction that there would be no large boulders or similar debris, already disproved by prior photographs, was given the boot when Armstrong and Aldrin walked among large boulders in craters near *Eagle.*

PLANETFALL

And the surface . . . Ah, at long last, to be able to tell by direct human observation *on* that surface. It wasn't sandy. It wasn't crunchy. It wasn't hard. It really couldn't be called dusty, although dust would be kicked up. It was described as powdery material, and a watching world saw the televised ghostly figures of two men kicking up spurts of that powder that fell in vacuum back to the surface. (For some reason the term dust became instantly popular, and the description of a powderlike surface material fell into disfavor. Dust it was, then.)

Dr. Fred L. Whipple had forecast that "loose dust on the lunar surface is practically nonexistent. . . . It should be sufficiently cemented or consolidated that it will not blow out dangerously in a rocket jet. To the human (encased) foot or under vehicles the surface should be 'crunchy' and allow minimal imprint."

Well, all *that* also had to be dumped and forgotten except for historical interest. The crew of Armstrong and Aldrin reported that while they were still some distance from the surface, fine dust blew up around the spacecraft and obscured their vision. (It would be much worse on some future flights.) And as for that minimal imprint, the pictures of Armstrong's boot, impressed deeply and perfectly into the lunar dust, became one of the great hallmarks of the mission.

But what did it *look* like? "It has a stark beauty all its own," Armstrong reported. "It's like much of the high desert in the United States."

There had always been worrisome questions about the ability of men, burdened with massive life-support equipment and their pressure suits, to move with proper balance and control on the lunar surface. The crew of Eagle walked, jumped and loped easily across the moon, bouncing in the one-sixth gravity and kicking up big clouds of dust that fell as quickly as they were lofted by the astronauts' boots.

On the morning of November 14, 1969, Pete Conrad, Dick Gordon and Alan Bean, the crew of Apollo XII, had an early breakfast with fellow astronauts and an unexpected guest—a stuffed gorilla in flight suit and crash helmet. Somberness was not the attitude of the second crew to depart for a landing on the moon.

Apollo XII launched in rain and climbed away from Earth amidst smashing blows of lightning and went on to sail to the moon. On the morning of November 19, the moon lander *Intrepid* dropped to the lunar surface in a blinding spray of dust. What Dr. Whipple had said would be "practically nonexistent" was so severe that Conrad was forced, as he explained later, to land in the blind—on instruments. All he could see of the moon in the final part of the descent was blinding dust streaming out from beneath his spacecraft.

Conrad and Bean, irrepressible in the light gravity of the moon, quickly dispelled some other predictions and dire warnings. Dr. Harold C. Urey had stated flatly that the requirements of landing on a precise spot on the moon were overwhelming, and he also said that the "chances of finding again the same spot on which the first landing was made are rather small. It would be necessary to spend years exploring the moon before it could be found."

Well, Conrad and Bean weren't looking for the abandoned landing stage of *Eagle*. Instead, they hoped to find something much smaller, the Surveyor III robot that had landed on the moon two *years* earlier. After five hours of work and preparation, Pete Conrad went down the ladder of *Intrepid*.

"They aren't kidding when they say things get dusty, whew!" he radioed. "I'm headed down the ladder. Man, is that a pretty looking sight, the LM."

The third man to set foot on the surface of the moon said, as his boot contacted the ground: "Whoopie! Man, that may have been a small one for Neil, but that's a long one for me. . . . Boy, you'll never believe it, guess what I see sitting on the side of the crater, the old Surveyor."

It was just 600 feet away.

PLANETFALL

Dr. Urey had also stated that if biological organisms were ever spread on the moon, then "within one month all bacteria would be dead . . . they would dry up completely."

Conrad and Bean brought back to Earth pieces of the Surveyor that had been on the moon for more than two years, exposed to vacuum, a temperature range of 500 degrees, solar storms, withering cosmic radiation. In the laboratory where those pieces were examined, scientists were stunned to discover a living organism. Not a lunar organism at all, but *streptococcus mitis,* deposited on the little spacecraft well before it was launched. It had survived the forces of launch, the weightless ride to the moon, the impact of landing, exposure on the moon for more than two years, the flight back, the reentry into our atmosphere— and it was *still* alive!

Armstrong and Aldrin proved that men could walk and even lope about on the moon. But what would happen if an astronaut, unbalanced by the heavy equipment backpack he needed to survive, were to trip and fall against the moon's surface? This was one of the more serious dangers anticipated before the manned flights. Well, Conrad tripped and fell several times. Instead of being helpless, Conrad found that simply by stiffening his arm suddenly—a dandy sort of one-armed push-up in the light lunar gravity—he bounded back to his feet. Other astronauts to follow, knowing that a trip or tumble wasn't dangerous, went through some rather violent physical gyrations in their work.

The question of dust rose again when Conrad and Bean worked steadily taking pictures, cutting pieces off Surveyor, collecting rocks and soil, and setting out their instruments. The dust in the Ocean of Storms, kicked up by movement, began to interfere with their work. It clung to their suits and their equipment, hampering photography. It was dragged back into their spaceship and created quite a problem before much of it could be removed.

Apollo XII gave scientists something else they needed urgently —comparisons between two landings. The lunar rocks returned by Conrad and Bean were surprisingly different in many ways from those brought back by the first mission. They had traces of potassium, uranium and thorium in greater abundance, with less titanium content, than was found on the Sea of Tranquillity. And the rocks astonished scientists by indications they were a billion years younger than those of Apollo XI. Conrad and Bean left behind them a nuclear-powered robot research station equipped with a seismometer, magnetometer, lunar atmosphere detector (for gases that might seep up from beneath the surface) and an ionosphere detector among other instruments.

After docking with Gordon in *Yankee Clipper,* the Apollo XII crew sent their ascent stage crashing into the moon's surface at 5,000 miles an hour. It struck 45 miles away from the research station they had left on the moon, and shocked scientists, who were studying the seismometer reports, with shock waves that clanged through the moon for almost an hour. A similar impact on Earth would have lasted about two minutes. Clearly the echo qualities of the lunar crust and interior demanded more study.

Apollo XIII never made it to the lunar surface; an explosion tore through the service module when it was nearly 200,000 miles away from Earth. In a cliff-hanging mission that was literally a life-or-death situation, XIII swung around the moon only 150 miles from the surface and was nudged by bursts of rocket flame back toward Earth, where it safely plunged back into atmosphere 142 hours, 40 minutes and 47 seconds after the flight began. It set one sort of record—it was the shortest manned round trip ever flown between Earth and moon.

Early in February 1971, the moonship *Antares,* with the Apollo XIV crew of Alan Shepard and Ed Mitchell, landed on the hilly terrain of Frau Mauro, and came to a stop with *Antares* resting on a slope of eight degrees. The crew would do more than just walk; they pulled a two-wheeled cart behind them to carry instruments and collect samples. Now men were climbing slopes and grades on the moon, proving even more strongly the ability of human explorers to enormously enhance scientific adventure. They carried out geological work, collected samples, took pictures, carried out scientific experiments, and moved within a field of boulders with individual chunks more than twelve feet high—providing overwhelming proof of violent impacts in the lunar past that had hurled house-sized debris across vast areas of the lunar surface.

Apollo XV ended all arguments as to whether or not a wheeled vehicle could travel safely and efficiently about the moon, when the crew of David Scott and Jim Irwin (setting the stage for the next two missions as well) bounced and jounced their way through dust, craters, up and down slopes, and through boulder fields in their electrically-powered Lunar Rover. The landing ship *Falcon* touched down on the moon on July 30, 1971, and Scott radioed: "Okay, Houston. Falcon is on the plain at Hadley."

It certainly was—smack in the middle of breath-taking views, since the Hadley Plain was a rolling region near Mare Imbrium, surrounded by towering mountains. The landing point was on the lunar plain called Palus Putredinis (Marsh of Decay), and the nearby Apennine Mountains towered 15,000 feet above them. They were close to the Hadley Rille, a trench depression 1,200 feet deep. In this single mission the crew could study three types of lunar topography: a mare basin, or plain; lunar rille, or gorge, and a mountain front.

"All the features around here are very smooth," Scott radioed. "The tops of the mountains are rounded off. There are no sharp, jagged peaks or no large boulders apparent. It's a gently rolling terrain."

The running commentary of Scott and Irwin, the views of Hadley Rille and nearby mountains, gave scientists valuable clues as to the composition and evolution of these geographic features. Ledges on the flank of Mount Hadley especially held interest as they seemed to be the lava marks of molten rock which flooded the adjacent basin. The rock formations dipping at an angle of thirty degrees, higher up the mountain, made it obvious these had been lifted up from the crust and then tilted. Nine layers were observed in the walls of Hadley Rille, providing a time-frozen picture showing that the plain extending to the west was formed by multiple lava flows over perhaps many millions of years. There had been no rocks larger than eight inches where *Falcon* landed. Near Spur Crater, one of their study points, the number and size of rocks increased dramatically the higher they went. The composition of the soil turned to especially fine powder. Huge boulders were found. Rocks of grey color and with a "thick, green colored layer" were picked up for samples.

Al Worden, orbiting in the command ship, later reported about the lunar dark side: "It is quite different from the front side. It is devoid of the large mare areas. It looks like continuous rolling hills that have been worked over a fantastic amount by meteor impacts. The surface has almost a fluffy look, as if it's been beaten and worked over a great deal."

By the flight of Apollo XVI, scientists had had a chance to put together the findings of the first three landings and they had come to certain conclusions. Most of the lunar surface history went back to a time when Earth's geologic record was murky and unreadable. The lunar maria covering a third of the visible face of the moon was made up largely of iron-rich volcanic rock that had come from a partially molten shell 100 to 200 miles beneath the surface. But within the mare basalts of Apollo XI and XII were intriguing fragments that clearly had not come from the mare, and had been deposited by violent origin some distance away.

Most of the northwest quadrant of the moon (near side, facing Earth) was of astonishly high uranium-rich volcanic rock formed about 4.4 to 4.5 billion years ago. Most highlands of the far side, and the eastern highlands of the near side, are of aluminum—and calcium—rich rock (anorthosite), which was first seen in the Apollo XI samples. Aluminum-rich material suggested that the primitive moon likely had a liquid outer shell 80 miles thick, which created the lunar crust simultaneously with the moon's formation. The old uranium-rich rocks limited the thickness of this early liquid lava and pointed to the composition of the solid interior very early in lunar history.

Scientists now deduced—and this would change from theory to solid belief—that they could eliminate certain theories regarding the origin of the moon. The chemical analyses showed the moon could not have been a body rushing through space that had been captured by Earth's gravity. Nor was it possible for the moon to have been fissioned off—ripped free—from an ancient Earth, from what is now the Pacific Ocean. The flights of Apollo had solidified knowledge and blown away the concealing mists of time.

It seems beyond all doubt now that the moon was formed from material which condensed, at temperatures much higher than previously assumed, out of the primitive dust cloud surrounding the early sun.

For the first time the beginnings of the solar system were emerging from the past. Long-standing and conflicting theories were being laid to rest, and the clues that indicated the origin of the moon enabled scientists to picture much more clearly the origins of the Earth. *Earth* science took enormous strides forward based on these studies of the moon—which was one of the earliest promises of Apollo.

By the close of the Apollo XV mission another astonishing find was confirmed. Three billion years ago the magnetic field of the moon was anywhere from a hundred to a thousand times stronger than it is today. We know that in eons past the moon rotated once every few hours and that it was much closer to Earth and that the effect of Earth's gravity forces were enormous on the moon. The mascons detected earlier with lunar satellite flights were refined as to detail and number. The massive matter concentrations were found beneath the five circular or ringed maria: Imbrium, Serenitatis, Crisium, Nectaris and Humorum. A sixth mascon was found between Sinus Aestuum and Sinus Medii and is suspected to represent an ancient mare obliterated by later surface violence. And right there was a critical clue—that what we saw on the moon was *not* the original surface frozen in time but a subsequent layer that was hiding the surface of the moon at

NASA Photo

The home planet of man as seen from lunar orbit during Christmas of 1968. It was the first detailed view of a known inhabitated world.

The extraordinary features of the Alpine Valley of the moon, looking west, as seen from a height of 153.5 miles. The surface characteristics are broadly similar to such features on Earth and Mars.

NASA Photo

NASA Photo

The mixture of thick dust-soil, scattered small and huge rocks at the Taurus-Littrow landing site of Apollo 17. The peak in center background is Family Mountain. Working on the Lunar Rover is Astronaut-Scientist Jack Schmitt, who spotted orange soil in the area.

The size of this huge split boulder at Taurus-Littrow is evident by comparison with Astronaut Jack Schmitt to its left, during the third EVA of Apollo 17.

NASA Photo

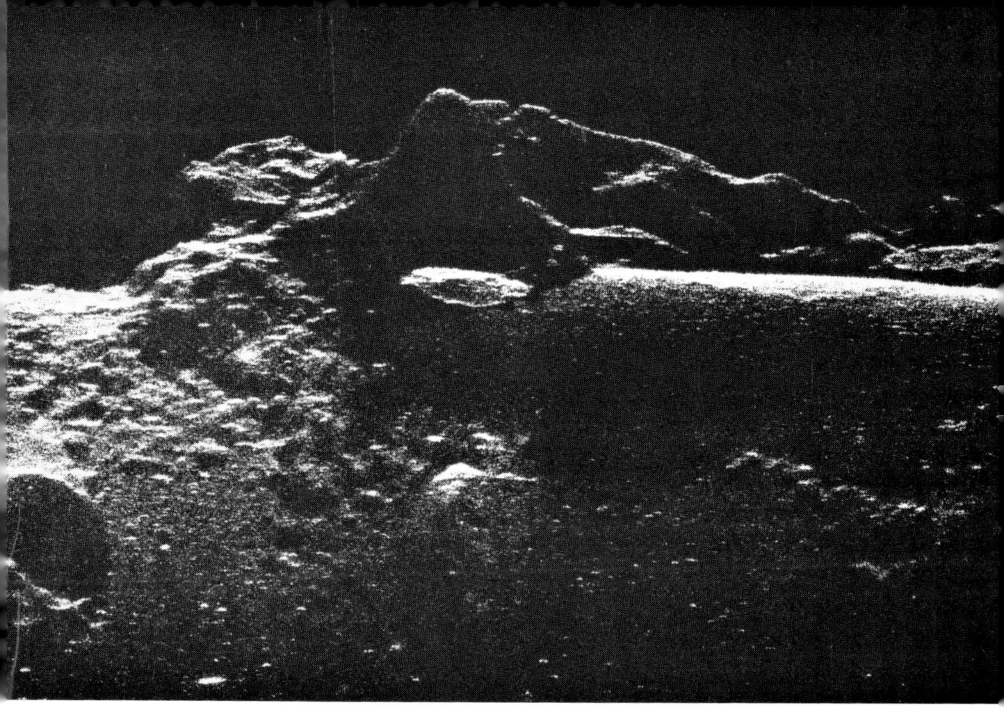

NASA Photo

The first closeup radar image of the lunar surface taken from Apollo 17 as it orbited the moon. The horizontal scale from left to right is just over 9 miles. Radar imagery is a great breakthrough in studying other celestial bodies. Here the mountains lift 8,000 feet above the basin floor. This first experiment is leading to a geologic model of the moon *to a mile beneath the lunar surface.*

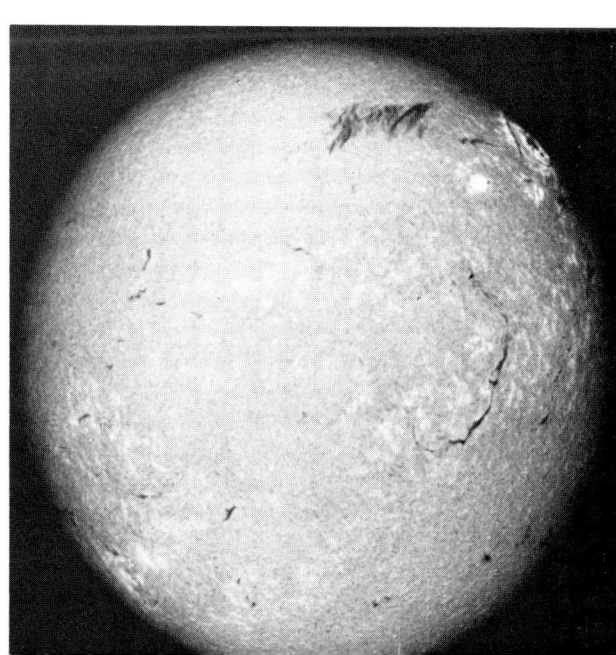

A vast storm on the sun (upper right of photo) covers an area 100,000 miles in diameter. The storm occurred in July of 1972 and set off wild fluctuations in the Earth's magnetic field, severely disrupting communications and power systems around the planet. Eight satellites tracking and measuring the event provided necessary information for scientists to start long-time predictions of such events.

NASA Photo

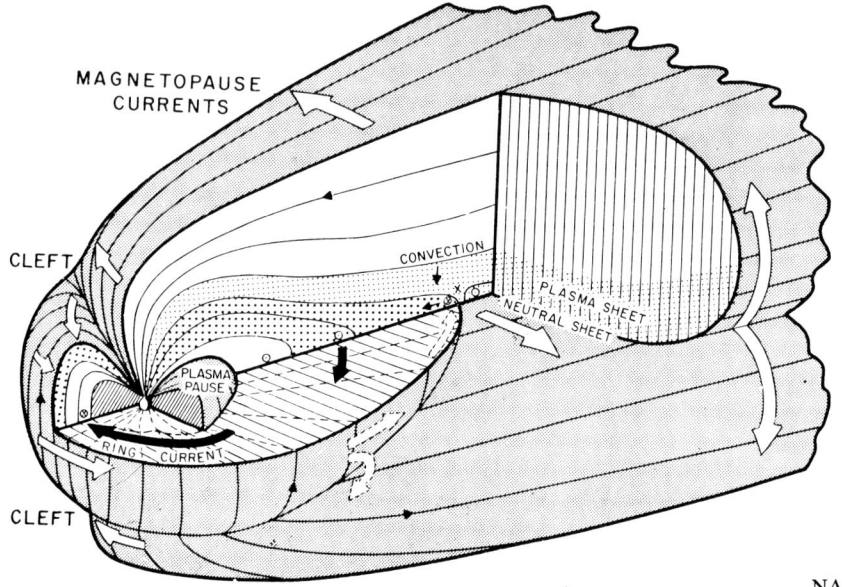

NASA Photo

A cross-section of the seething violence that surrounds the Earth (seen as a small circle in left center). Satellites like the IMP (Interplanetary Monitoring Platform) orbiting halfway between Earth and the moon's orbit have charted such energy fields and their changes in great detail.

Objectives for Tomorrow—from Mercury outward to Pluto. Shown here are the comparative sizes of the planets and the sun; the planetary orbits are not to scale. Note the sharp deviation from the plane of the ecliptic of Pluto, sometimes called a rogue planet.

NASA Photo

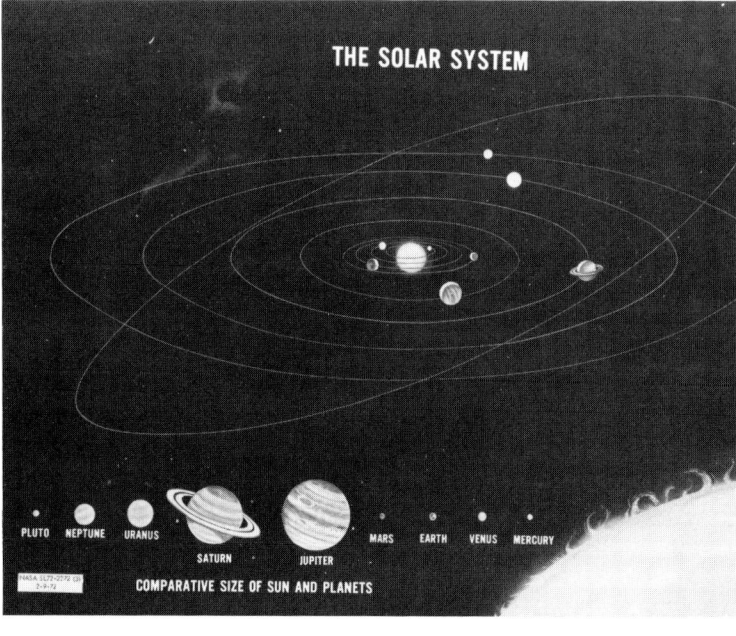

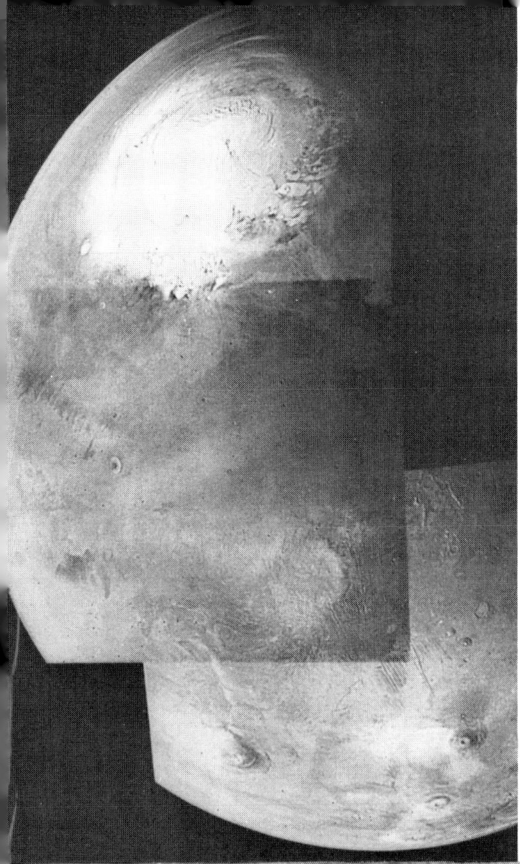

NASA Photo

One of the most extraordinary of the photos of Mars, taken from an average range of 8,500 miles, and showing the north hemisphere from the polar cap to a few degrees south of the equator. The polar cap (top) is shrinking during the late Martian spring. In the center of the disc can be seen fractured terrains partially flooded by volcanic extrusions. But the great volcanic features, extraordinary in their clarity from space, dominate the globe. The volcano in the lower left is Nix Olympica, 310 miles in diameter and partially obscured by clouds.

The greatest volcano known on at least three worlds—Nix Olympica—shows a mountain bigger than the state of Missouri. The greatest volcanic pile on Earth, in the Hawaiian Islands, is less than half the size of Nix Olympica. The main crater at the summit, a complex multiple volcanic vent, is 40 miles in diameter.

NASA Photo

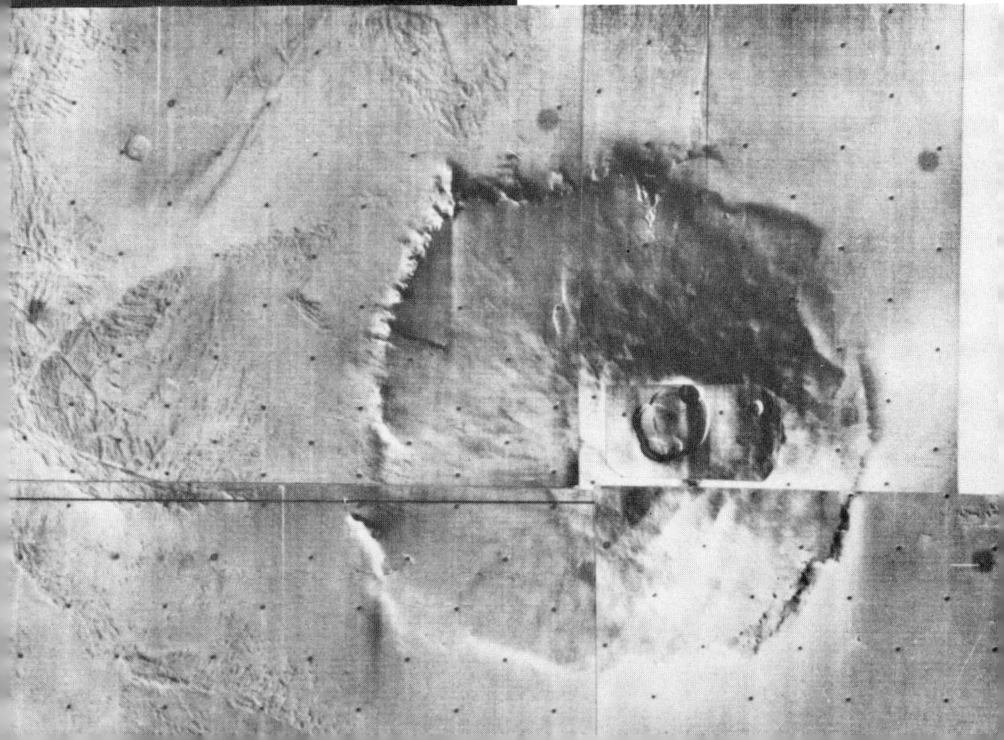

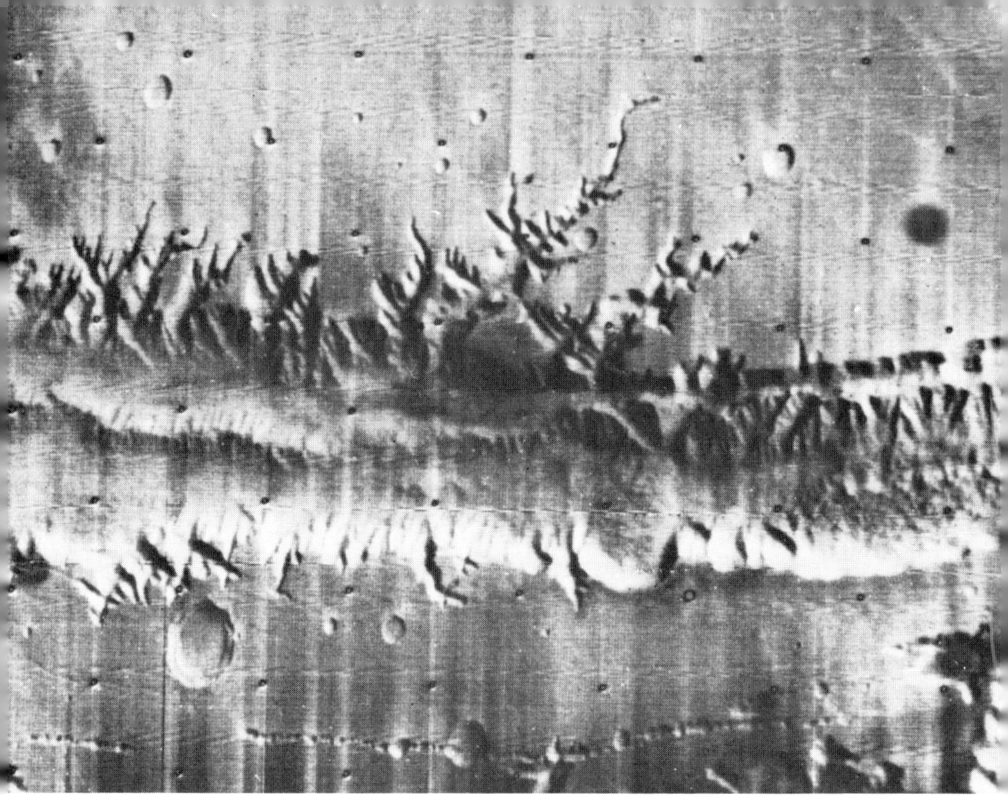

NASA Photo

A vast chasm with branching canyons eroding adjacent plateau lands in Tithonius Lacus, 300 miles south of the Martian equator. The photo was taken 1,225 miles above Mars and covers 235 x 300 miles.

Compare this photo with the previous and following pictures—but this one is 568 miles above the Grand Canyon in the U.S. The canyon bears an incredible resemblance to similar, but vastly larger, features on Mars. In far left center is Lake Mead, and the Grand Canyon wanders to the right across the photo to upper right.

NASA Photo

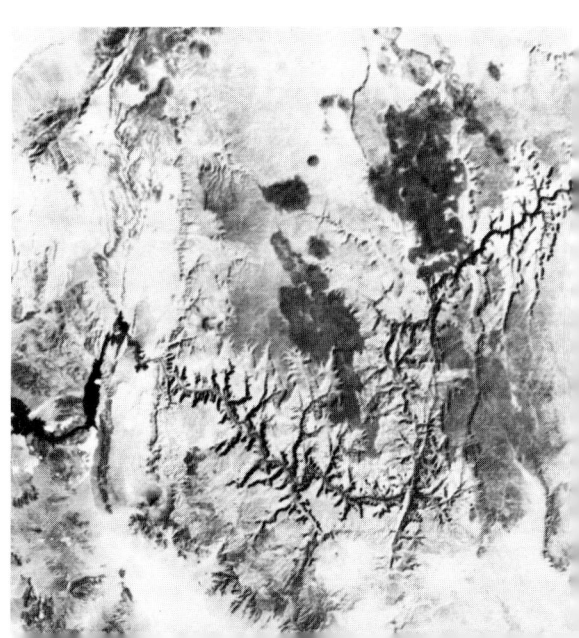

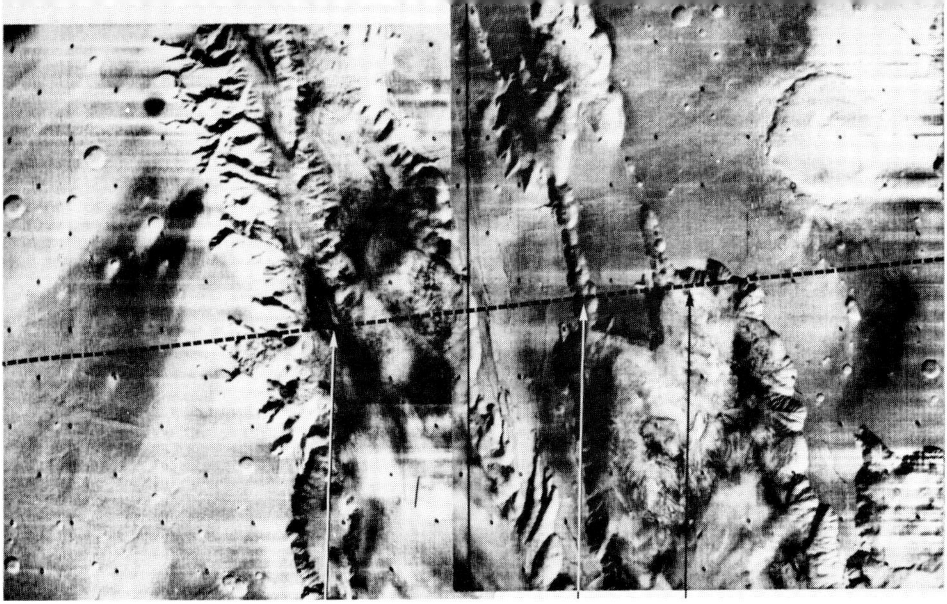

NASA Photo

A mosaic of the massive canyon of Mars from 1,070 miles altitude. The white arrow at left points to the greatest depth; when this picture was first studied the depth was estimated at 9,500 feet, but that figure has since been refined to 20,000 feet—nearly four times as deep as the Grand Canyon in the U.S. The photo covers an area 400 miles across. The Martian canyon is 75 miles wide compared to the Grand Canyon at 13 miles.

At right is a dune field 80 x 40 miles in the Hellespontus region of Mars. The white arrow points to a single crater 93 miles wide, and the enlargement of this crater (left) shows many long dunes spaced about one mile apart. The dunes were created by strong winds blowing consistently from the southwest.

NASA Photo

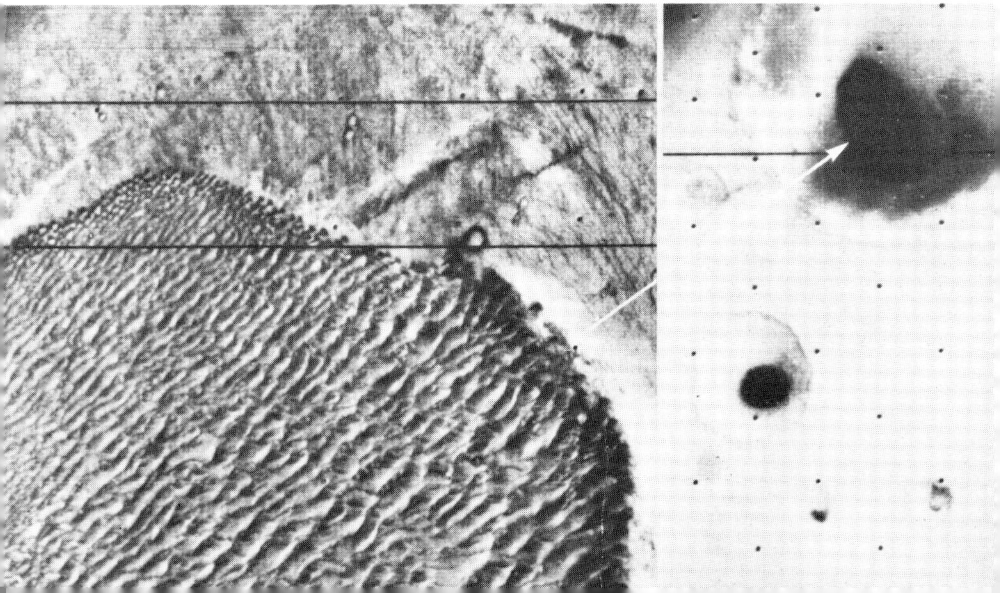

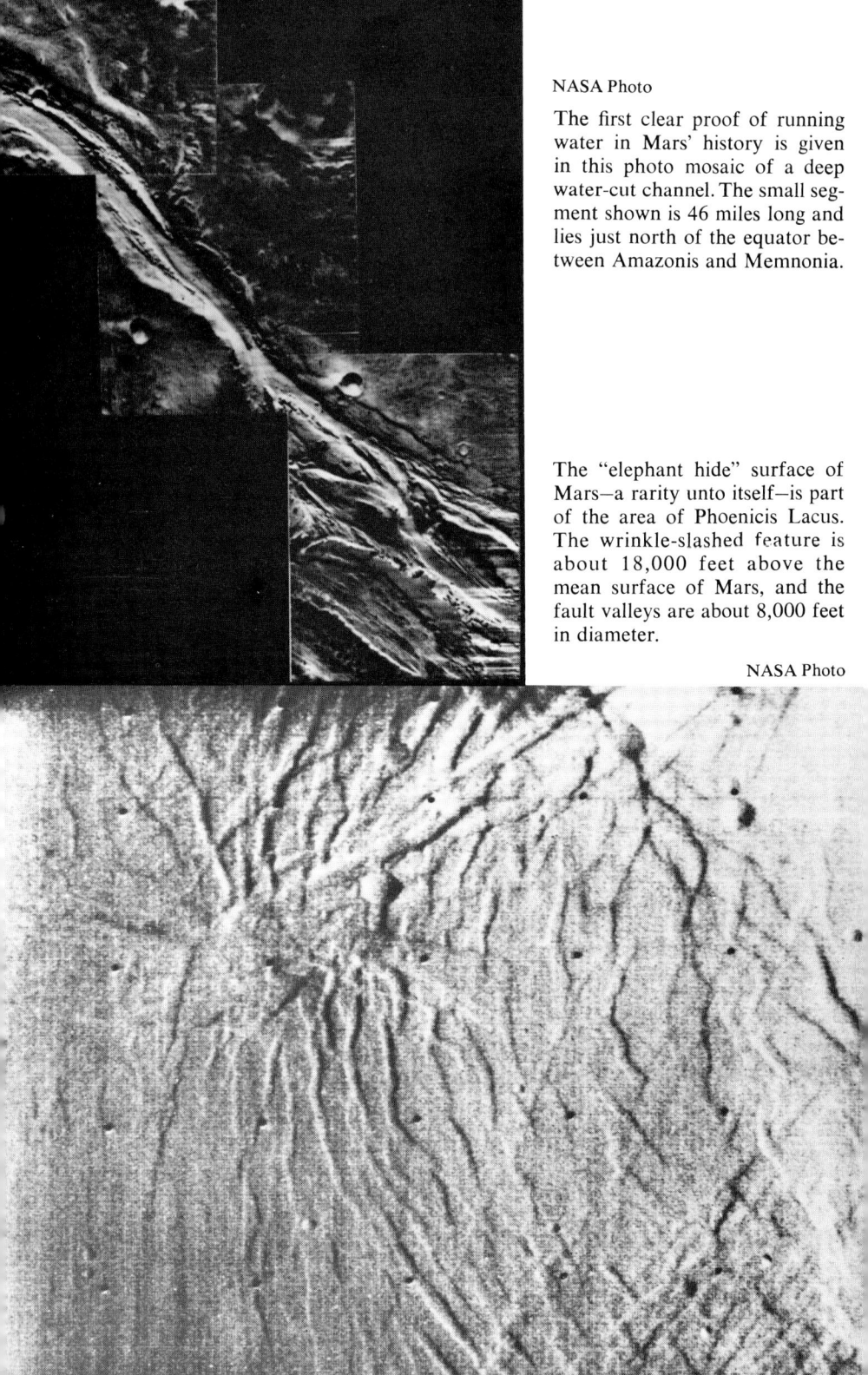

NASA Photo

The first clear proof of running water in Mars' history is given in this photo mosaic of a deep water-cut channel. The small segment shown is 46 miles long and lies just north of the equator between Amazonis and Memnonia.

The "elephant hide" surface of Mars—a rarity unto itself—is part of the area of Phoenicis Lacus. The wrinkle-slashed feature is about 18,000 feet above the mean surface of Mars, and the fault valleys are about 8,000 feet in diameter.

NASA Photo

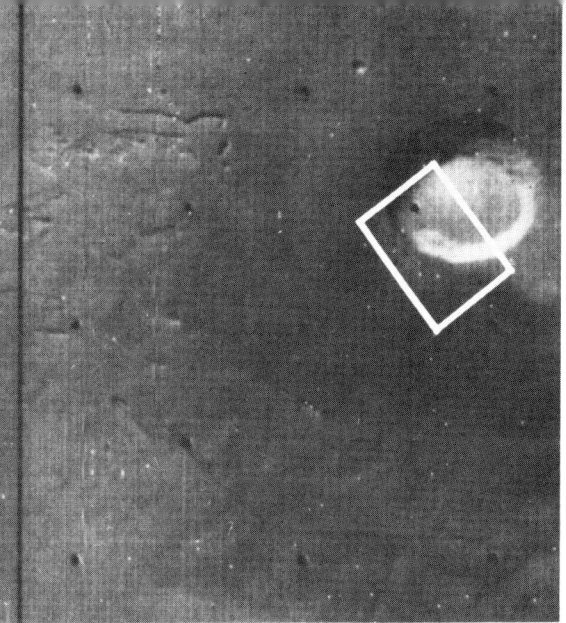 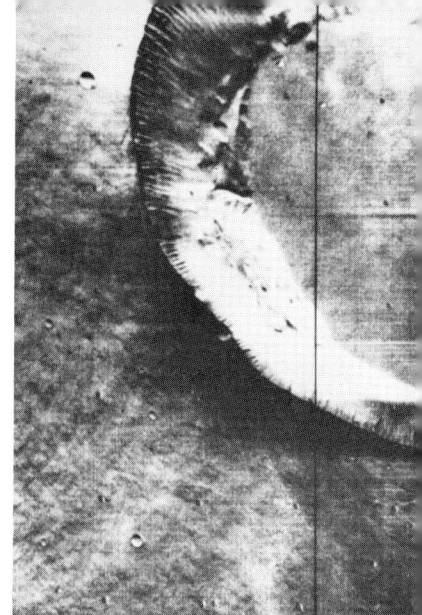

NASA Photo

A great Martian shield volcano called Middle Spot, photographed from a height of 1,209 miles. The "flat" picture was taken with a wide-angle lens, and the picture in great detail with a telephoto lens. This shows the crater, 25 miles wide. The rough volcanic flank is splattered with impact craters. The smooth crater floor is a former lava lake.

What appears to be an uncovered ancient metropolis on Earth is really a natural geometric pattern near Mars' South Pole, made up of a complex of transecting ridges. The area shown is 26 x 30 miles.

NASA Photo

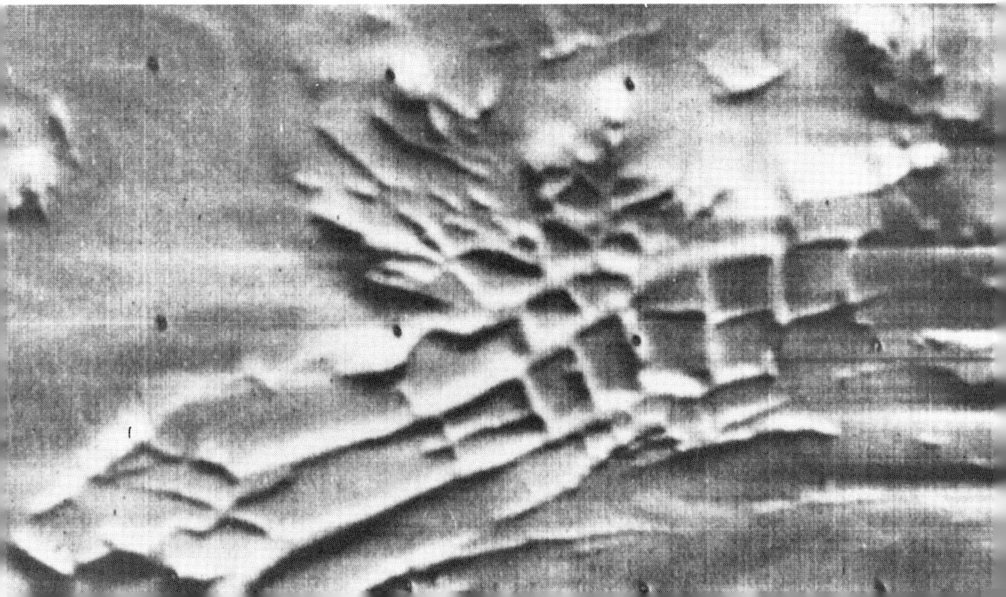

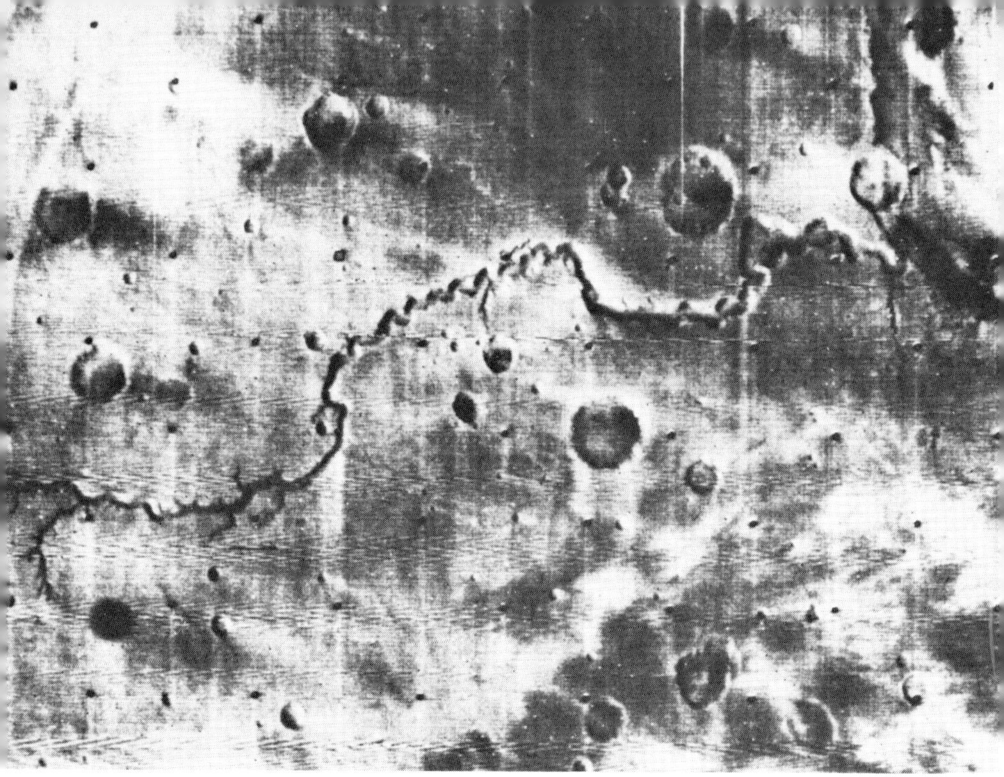

NASA Photo

A sinuous valley of Mars from 1,033 miles. The valley is 250 miles long and 3 miles wide and resembles a giant version of an Earth *arroyo,* common to the southwestern U.S.

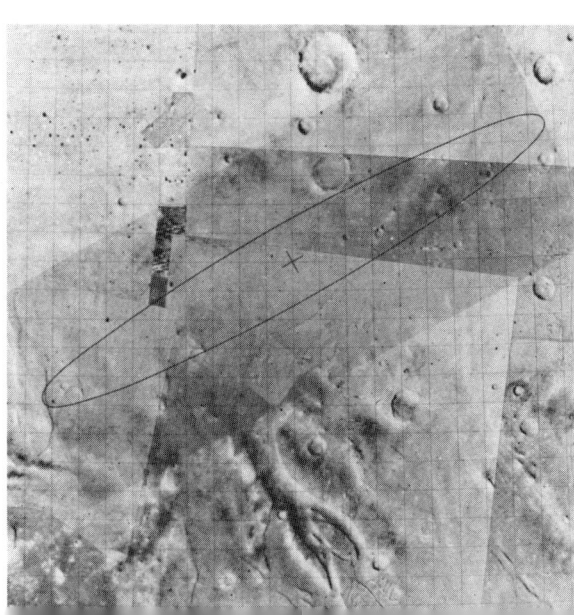

The area shown in this photo mosaic of Mars, north of the 3,000-mile rift canyon along the equator, is Chryse—the Land of Gold. The sweeping oval depicts the planned landing site of the first Viking Lander, scheduled to touch down on Mars on July 4, 1976. The oval—or landing ellipse—covers 62 by 370 miles. Scientists state the air pressure at Chryse will sustain free water vapor and high enough temperatures to support life.

NASA Photo

NASA Photo

The Viking Lander depicted after a successful landing in the Chryse region of Mars. A surface scoop to bring Martian soil to instruments within the Lander is shown making its third dig.

Why we need so urgently to obtain close-up photos of the planets. Shown here are among the best Earth-taken pictures of Mars, Jupiter, Saturn and Pluto (not to scale). Look again at the photos of Mars taken by the Mariner spacecraft!

TRW SYSTEMS

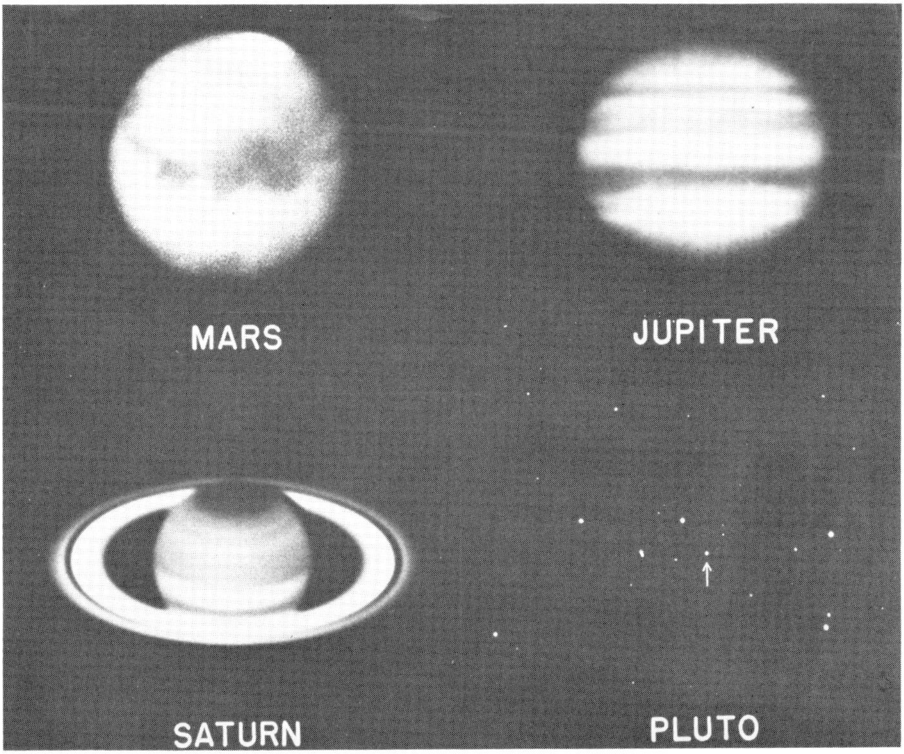

NASA Photo

The moment of launch for the TRW Pioneer XI aboard an Atlas-Centaur from Complex 36 at Cape Kennedy—9:11 P.M., EST on April 5, 1973. The second TRW Pioneer spacecraft sent toward Jupiter is destined to arrive at the great planet in December 1974, one year after the arrival of TRW's Pioneer X.

Huge space-tracking antenna in Australia, near Canberra, to support NASA's deep-space exploration program, starting with Pioneers X and XI.

NASA Photo

NASA Photo

A stylized view of Earth, moon and Pioneer X—launched from the Earth with a speed of nearly 32,000 MPH. Four and a half years after launch it will cross the orbit of Saturn. And, 9 years after launch, 1.8 billion miles from the sun, it will cross the orbit of Pluto— and begin the first journey to another star.

A representative view of Pioneer X and XI (known as Pioneers F and G before launch) missions to Jupiter, listing the major scientific experiments.

NASA Photo

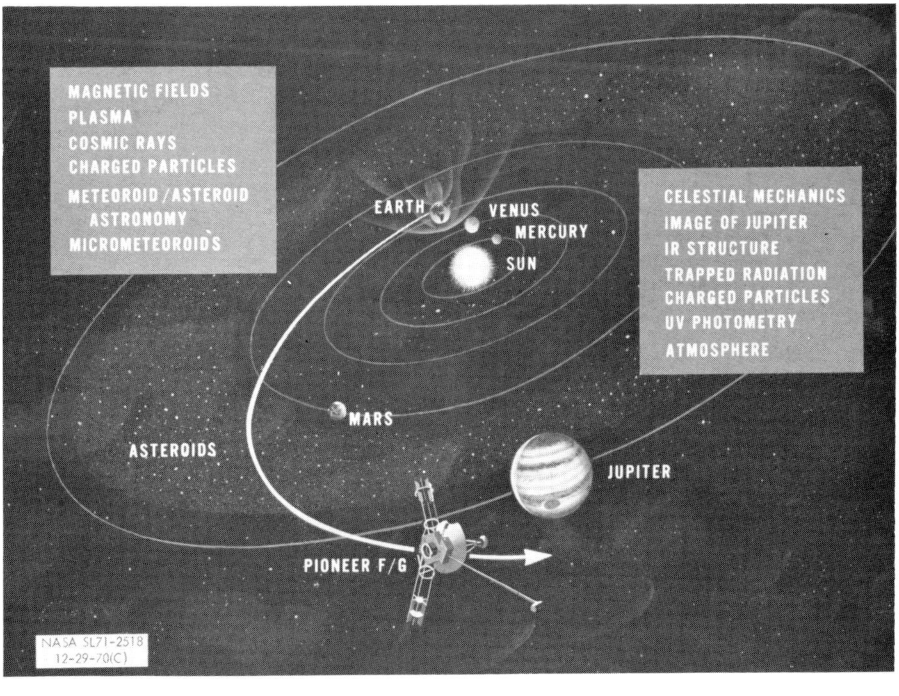

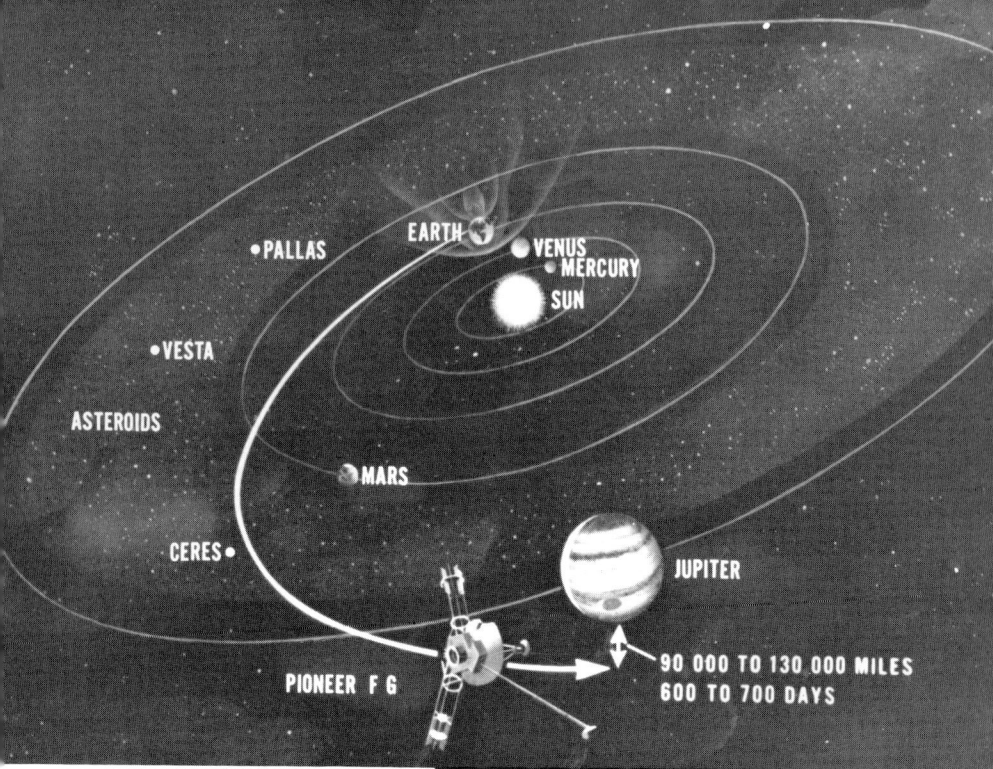

NASA Photo

A representative view of the orbital path of the Pioneers X and XI spacecraft, the first robot explorers to cross through the dreaded asteroid belt (Ceres, Vesta and Pallas are indicated) en route to Jupiter. By midsummer 1973 Pioneer X (Pioneer F) had successfully completed that part of its enormous journey.

An outstanding Earth-based telescope view of Jupiter, clearly showing the enigmatic Great Red Spot (upper left).

TRW SYSTEMS

NASA Photo

Artist's portrayal of Jupiter and its four larger moons. From left to right: Ganymede, Europa, Io, and Callisto, all with orbits lying between 262,000 and 1,117,000 miles from the massive planet. Jupiter has 12 moons in all.

A dramatic and beautiful impression of the Great Red Spot on Jupiter, as Pioneer X hurtles by the planet on its way beyond the solar system.

TRW SYSTEMS

TRW SYSTEMS

The beauty of Saturn as seen by an Earth-based telescopic photo. First visitor to the great ringed planet beyond Jupiter is expected to be TRW's Pioneer XI, now scheduled to pass by Jupiter in December 1974.

TRW's Pioneer XI in an artistic portrayal as it sails past Saturn.

TRW SYSTEMS

the moment it solidified. The mascons were also identified as circular plates of dense rocks filling the lunar mare, formed originally by the collision of small mountain-sized asteroids against the lunar surface. And then another six mascons, for a total of twelve, were discovered beneath the surface of the lunar far side.

Another clue to Earth's history was found in studies of Apollo XIV material. The largest collision of an asteroidal object likely took place four billion years ago, which is much later than previously estimated. The importance closer to home? The revised picture of the early days of the solar system shows that major collisions of mountain-sized asteroids against the Earth were taking place for more than a billion years after our planet was formed—and opened the door to more accurate study of Earth's past history.

Apollo XVI, launched by a Saturn V that weighed more than 3,500 tons, carried John Young and Charles Duke to the moon aboard their landing ship Orion on April 20, 1972, with touchdown in the Descartes region. Instead of the volcanic rocks scientists expected they found breccias—aggregates of many materials welded together in a single rock—which could have come from volcanic eruption *or,* more likely, from the explosion of a large meteorite which broke off the surface material and remelted it before it again solidified. They picked up crystalline rocks never seen before. They turned over boulders and took samples of lunar soil covered by the rock for billions of years.

XVI proved that the surface of Cayley and Descartes Formations, where they landed, are ejecta flows: tremendous streams of molten material thrown out by meteor impacts that surged across the moon's surface for long distances. The rocks that formed in the wake of these streams as the material cooled are rich in anorthosite, the earliest known rock found on the moon, which was suspected to be part of the original crust. High along the flank of Stone Mountain, Young and Duke studied huge blocks on the side of South Ray Crater once tossed out from the impact crater. "Fresh" meteor craters were filled with breccia containing the purest anorthosite yet found. An idea of the size of the boulders is given from one they named House Rock, because it had been tossed some unimaginable distance, and was still the size of a three-story house.

Challenger, the landing ship of Apollo XVII, took Gene Cernan and Jack Schmitt on December 11, 1972, to the high mountain country of Taurus Littrow for the last mission of Project Apollo. There were hopes that the final team would bring back lunar rocks considerably older than the most ancient samples—4.25 billion years—picked up during Apollo XVI. But nowhere in the 249 pounds of rocks was the elusive historical lodestone to be found, and the very absence of this dated specimen helped cement some new theories about the moon. Schmitt had discovered orange soil by a lunar crater, and this was the other end of the time spectrum; there were hopes the orange moon dust would be evidence of recent volcanic activity. Thus the search for the oldest, and the youngest, rocks was not to be answered, and the last samples returned from the moon fit neatly within the time brackets of earlier missions.

The orange dust, in fact, turned out to be tiny glass beads created by a meteor slamming into the moon some 3.71 billion years ago, instantly melting itself and the lunar rock it struck so violently. The explosion sprayed the melted liquid out into space as very tiny droplets, traveling at high speed and spinning rapidly. Most of the droplets cooled very fast, forming solid orange glass beads before they fell back onto the moon. Others cooled more slowly, forming crystals and turning black. The orange samples derived their color from a high titanium and iron content.

Apollo ended with one particular mystery left unsolved. Volcanic lava samples were returned from the moon in great quantity, but no astronaut was ever able to visit a clearly identified lunar volcano. This very omission again shifted the time scale of events on the moon, and also reemphasized the conviction that the original lunar surface had been overlaid with material deposited since the crust cooled.

Little matter against the staggering revelations that changed forever the murky groping back in time. The moon emerged from the chemical analysis of its materials as ancient as the Earth. It is fully a planet in its own right, and the history of the moon is the history of the entire solar system.

Until Apollo came to its end, the most widely accepted estimates for the age of our solar system included a figure of six billion years. Gone forever, that number, for research on the rocky, dusty moon has made it clear that it was about 4.5 billion years ago that the sun and the planets were formed from the same mixture of gas, dust and ice. Because temperatures were so high close to the sun, Earth and its moon, Mars, Venus and Mercury are comprised of rocky material that condenses at high temperatures, while ices and gases are largely blown away.

Far from the sun it was cold enough for water, methane and ammonia to remain as ice and gases, and so the outer planets and their satellites are formed of these elements.

Shortly after the moon was created—when it was what scientists call an "infant world"—it went through a savage battering by billions of meteorites and asteroids, some of them, as we have noted, "mountain sized." The impacts were so violent that massive chunks of rock were hurled into space to tear through the thick atmosphere of a primitive Earth in the form of thundering, flaming debris. One large impact, the event that produced the Imbrium Crater, was so violent it shattered open a vast section of the lunar crust and hurled a blanket of debris that in itself was mountain sized in some areas of the moon. It's difficult to think of entire mountains originating from such impacts rather than being uplifted from within the crust, but there's no question any longer that it happened this way. The impact crater known as Oriental could swallow the entire state of Texas, and its formation was so extreme that some five percent of the mass of the exploding meteorite and the melted moon materials was flung into space so violently it exceeded the escape velocity of the moon (1.5 miles a second) and fell toward Earth—creating what was likely the most stupendous meteor shower in the history of the planet we call home.

Such global catastrophe began to explain why the original crust of the moon was never found during the manned explorations of Apollo. The moon, as the story emerges, was formed 4.5 billion years ago. Over a period of a few hundred million years, the molten outer layer of the moon solidified—the layer we sought during Apollo. But about four billion years ago there began the cataclysmic impacts of such extraordinary power that the primitive lunar crust was melted, churned up, destroyed and buried beneath the debris of the massive explosions.

It has been this history, drawn painfully from Apollo and other research, that finally gave to us here on Earth the first accurate signs of how our own world was born.

CHAPTER 10

Sister World

Everything you ever heard about it is true. This business of looking at Earth from lunar orbit. There's no way ever to capture the stunning sight on film. Not on pictures or on sensitive motion picture film. No matter how sharp or realistic that film is it's still flat. And the home world of man from a quarter-million miles across vacuum is not a flat still life. An artist, well, that's another matter. An artist has the freedom to paint what he sees.

Ever since men first journeyed to the moon they have tried to tell us how pale and washed out were the photographs they brought back with them. One evening—long into the night, actually—I sat with Gene Cernan and Ron Evans of Apollo XVII. Gene and I were talking about the future of the human race, how there could never be a separation again of man and his responding to the instinct, the absolute need, to travel far from the world of his birth. To deny man this physical and spiritual soaring, once we knew the wonders and the knowledge awaiting us, would be the same as amputating the wings of a falcon. It would live, but it would no longer be truly alive.

What Gene, who also flew Apollo X into lunar orbit, tried to convey was the stunning beauty of seeing Earth from so far across space. Whatever I say here and now, trying to capture Gene's words, almost certainly will be an injustice to those remarkable moments he shared with us. For Gene Cernan is a deep and emotional man, just as much as he is a brilliant, sharp, intelligent, skilled—and courageous—craftsman of his trade, which is plying the seas of space between worlds.

PLANETFALL

No man, said Gene, could ever be the same within himself, or in terms of his outlook upon mankind and life, after being exposed to an awesome tapestry painted across the fabric of space itself. Coming up and around the moon in lunar orbit the sight of the distant Earth is startling. First, there is that utter black of space. It is a complete, overwhelming, utterly overpowering black upon black upon black. There are no stars to be seen at this moment, for they are in sunlight, and adaptation of the eye has yet to take place. And there, those quarter million miles away, against this most velvet of blacks and black of velvets, rolls a spherical jewel, of a stunning blue; whorls and tendrils of white of her clouds; and where there is the terminator, the line between night and day, the atmosphere, the clouds, the rolling curving edge of the planet shows its dark red tendril that bands the home of man. Above all, Gene stressed, was the sensing of the *sphere*, the roundness that can never be captured on the flatness of two-dimensional film. It was as if, said this man who has flown twice from Earth to the moon, you could actually see around the curving edge of that distant, small, stunning world, and you knew, above all, that this was *real*.

This is how God would see the world of men . . .

That was long ago, and Gene Cernan was the last man of Apollo to walk the surface of the moon. Now, we are ready to leave the moon of 1973 and return to that future time of 1998, aboard the nuclear drive spaceship *L. Gordon Cooper*. Fueling has been completed. The power grids slide upward through vacuum the huge containers with their liquid fuel for the nuclear generators. Everything is in readiness. The *L. Gordon Cooper* is an exquisite timepiece of millions of parts all wound to a fine pitch, in harmony with the balancing of the celestial spheres, ready to obey the laws of Copernicus and Kepler and Newton and Einstein, those "giants" of whom Frank Borman spoke when he returned to Earth after the first voyage to the moon.

But this ship with its mighty nuclear engines uses the moon—

SISTER WORLD

or the orbital path about the battered planet—as a way stop on the curving celestial highway to other worlds. Our destination, this time, is the sister world of Earth, the second planet outward from the sun, the celestial siren, forever, as man knows this word, concealed beneath and behind shrouding mists that reflect light and create a glowing, brilliant jewel in Earth's night skies.

For two days in lunar orbit we have been weightless. Now, that ends. No need to strap in this time, for the nuclear drive comes alive gently, at only a fraction of its full thrust, and the giant spaceship accelerates slowly but inexorably, increasing the distance between the lunar surface and its gleaming assembly. There is no sensation of speed, there is only the mild acceleration that provides a comfortable partial gravity within the ship. But sooner than we expect the invisible chains of lunar gravity are severed. Not long after the velocity has increased to the point where Earth no longer holds its massive sway. *L. Gordon Cooper,* if all power ended abruptly now, would never return to Earth. We are in solar orbit, a man-made asteroid. But only for a short time.

Well away from Earth the engines increase their thrust. There is acceleration for a short time that is solid, almost a full g-force, almost the same as standing on the Earth's surface. But *L. Gordon Cooper* is not increasing its speed—relative to the sun. While the engines pound with their muted thunder, hurling the intense violet glow *in the direction of flight,* the spaceship decelerates. It must reduce its orbital speed around the sun so that when the engines shut down, we will be accelerating under solar gravity. Our speed increases again but now from gravitational energy of the sun, and not from our own power.

The laws are immutable and they can be made to work for us. A satellite in orbit about the Earth races over the world with a speed of 17,500 miles an hour. It is in a perfect balance between the centrifugal force of the rocket thrust that hurled it up and outward from Earth and the gravitational attraction of the planet below. If we increase the speed of that satellite from its

PLANETFALL

orbit, it moves away from the planet—outward bound. This is how we flew our Apollo missions to the moon. First, orbit; then, a long burst of power to increase speed relative to the Earth.

To deorbit, to return from orbit at 17,500 miles an hour, the trick is to do the opposite. Again you expend energy, but this time in the direction of your flight. This slows you down to, let's say, 17,100 miles an hour. That slight difference is everything. Gravity is now the master of centrifugal force, and Earth reclaims its own. You're on the way down.

A spaceship in the vicinity of the Earth (which includes the moon) moves about the sun with the planet's orbital speed of 66,000 miles an hour. If we wish to move outbound we increase our velocity (along the plane of the ecliptic; remember?), and we're on our way to a meeting with Mars or Jupiter (depending upon the details of expending energy and in what direction).

If we wish to get closer to the sun—the ruling gravitational body—we've got to do precisely what a satellite must do to leave orbit and return to Earth. Relative to that primary body about which we orbit we must slow down.

This is what *L. Gordon Cooper* has been doing—firing its engines along its direction of flight so that, in terms of orbital speed around the sun, we lose speed. When the engines shut down we're on an inward curve from Earth's orbit toward the sun. But because we know our exact position and velocity in space relative to the sun, and we know the exact position, movement and velocity of Venus, we can time our sunward fall so that our path through space will intersect the path through space of Venus. This is a rendezvous on a grand scale, but that's the road map we follow. We have one enormous advantage over the older chemical rockets. With so much energy at our disposal we can thrust for long periods and reduce the travel time between worlds. Instead of many months we can count our flight in terms of weeks.

When the chemical barges (which is what the astronauts of 1998 call the old rockets like the Saturn) of the early days of

SISTER WORLD

space flight were used between Earth and moon, power was so scarce (sixty pounds of fuel for one pound of payload to the moon) that the most conservative flight regime had to be followed. It took Apollo about three days to travel from Earth to the moon, and, at the neutral point between the two worlds, the spacecraft was almost crawling through vacuum, so low was its speed.

But what if there had been a surplus of power? Apollo could have accelerated to 29,000 miles an hour instead of 24,000 and cut the trip down to perhaps a day. Of course, there would have been a far greater energy expenditure to get into lunar orbit because of all that excess speed. You see the point: get to the moon with the greatest possible payload, which means the lowest possible speed, which in turn demands a longer flight.

The first robot mission to Jupiter left Earth with a speed of 32,000 miles an hour. Apollo at 24,000 miles an hour needed three days to reach the moon. The Jupiter probe made the same distance in eleven hours.

Now, in this huge nuclear-drive ship, we have power to spare, and we can exchange energy for speed and a flight of greatly reduced duration. Which is all to the good. Beautiful as may be its splendors, spaceflight is essentially a boring way to get from here to there. There's no motion, the sightseeing is almost always the same, and the moments of utter beauty must be taken in small doses. We appreciate a sunset because it's here only for moments. How long would you keep looking at a sunset if it lasted for three months? How many times, without stopping, would you listen to your favorite symphony before you had the urge to tear the place apart? On the return from the second manned flight to the moon, Pete Conrad made this point. The three-day return trip was essentially boring, and Pete sounded a call for future flights to carry music tapes, reading materials, *anything* to avoid the repetitive sameness.

* * *

PLANETFALL

Well, at least you have plenty of time to think about where you're going, and that moment when the crew slams on the nuclear brakes to sling the great ship into high orbit about Venus.

Strange, when you think about it, how a lack of information so often grows by leaps and bounds into a belief that has no scientific basis but becomes "accepted fact" simply because enough people *want* to believe it. Few things irk men of science (and they aren't all *that* honest) more than having to respond to questions with a puzzled look on their faces and a collective shrug of shoulders. People have a nasty habit of assuming that scientists should know about those matters on which they're questioned—if for no other reason than that scientists spread this belief and spend great sums of money collecting information.

But with all his instruments and a lifetime of study, the scientist doesn't really have the faintest idea of what it may be like on Venus. Oh, he's got ideas (most of them horribly wrong), *but he does not know.*

If you don't know—at least *say something.* Don't *quite* make it up out of thin air. Deduce. If you have only a shred of cloth, weave yourself a magnificent set of clothes by mixing liberal amounts of imagination with that shred.

That's just about what happened with what we *thought* we knew about Venus. The theories were both serious and preposterous. They were sincere and they were outlandish. They were well intended and they were based on everything we knew about Venus, but people couldn't separate minimum fact from maximum imagination, and what emerged was gibberish.

There's an old foot-in-the-mouth story about this sort of scientific thinking that says: Hundreds of scientific experiments have proved conclusively that the beating of tom-toms always causes the sun to reappear after an eclipse.

Hard to argue that when the natives have been beating those drums for hundreds of years. They're not taking any chances. They *know* that the drumming always brings the sun back, because every time there was an eclipse and the tom-toms thun-

SISTER WORLD

dered, the sun *did* come back. And they're just not going to take any chances, because they can't afford not to have the sun around every day.

The fact that there can't be any causative relationship between whaling away at the drums and the coincidental positioning of sun, moon and Earth just isn't brought into the matter.

This brings us back to Venus, which we said earlier was the Sister World of Earth. And it is—*in certain ways.*

There are the basics of the two planets. Venus is the planetary body most like Earth in general physical characteristics. Venus's diameter of just over 7,700 miles is only 200 miles less than that of our world, and the difference is razor-thin. So thin that the expression sister planets is grounded very solidly through this one similarity alone. Of all the planets in our solar system the most dense is Earth, but Venus comes close, with a surface gravity eighty-seven percent that of Earth, which means that a man who weighs 200 pounds (terrestrial pounds) would, on Venus, tip the scales at 174 Earth pounds.

Well, there's more than that, of course. Except for the moon, Venus is the brightest object in the night sky, and there has always been the feeling that there *must* be a great similarity between Venus and Earth because of the thick atmosphere on both worlds. It turns out that Earth is a planet highly reflective of sunlight striking its surface—but in this case, surface must include ice caps, clouds, snow fields and the atmosphere, as well as special effects such as the sun creating dazzling highlight spots across the oceans. Venus, however, is even brighter from space than is Earth, since the light it reflects (albedo) is just about the highest of any planet in the solar system; the figure is seventy-six percent of all light bounced back.

Venus rolls through its solar orbit for a complete revolution once every 224.7 Earth days. For a long time—until very recently, in fact—scientists were convinced that Venus duplicated about the sun the motion of the moon about the Earth, in that the period of rotation and revolution matched closely. Thus

PLANETFALL

Venus would present, always, the same side of the planet to the sun, just over sixty-seven million miles away. The combination of absolutely *not knowing* the characteristics of the Venusian atmosphere, coupled with this unusual performance, brought forth some of the most lurid descriptions of a planet ever known along the periphery of astronomy. And at times the slim dividing line was crossed, as scientists, pressed for details of this enigmatic world that comes as close as 25 million miles to us, slipped, from a lack of facts, to hyperactive speculation. Nothing wrong with speculation, but it has a habit (with no fault to the source) of being whipped into an exciting froth.

Then, through radar studies of Venus from Earth, another strange fact emerged. All previous information about the rotation of Venus was incorrect. Radar scanning showed, this time with little question, that Venus rotated not once every 230 days (to closely match the period of revolution) but once every 243 days. This meant, of course, that the same side of the planet did not always face the sun, although each day and night, by our standards, was distressingly long. Down in Puerto Rico there is a radio telescope (the Arecibo Ionospheric Laboratory) a thousand feet in diameter that stayed glued to Venus to extract the painfully few facts about the shrouded globe. The Arecibo scientists showed not only that Venus rotated once every 243 days but that it rotated in a clockwise direction—opposite that of every other planet. As if to add insult to injury in terms of distinction, Venus is also the planet with an orbit closer to a perfect circle than any other body in space. And, with its clockwise direction, an observer on the surface of Venus (who could hardly see at all because of the massive atmosphere) would, if he were patient, see the sun rise once every 117 days (Earth time)—in the *west*.

Well, back to that thick Venusian atmosphere and the question that haunted scientists and gave rise to the lurid tales of what went on beneath its blanketing cover. Before the first probes reached the planet, when we knew even less about Venus than we do now, it was acceptable to speculate about the atmosphere, which is another way of saying you're guessing, but with fancy words. Well, Earth and Venus are of similar size, mass and surface gravity. Both planets are bound by heavy atmosphere. Both planets—so went hard scientific conviction—have been "cooling off" from eons past when temperatures were much higher than they are today.

Now, this led to an extraordinary conclusion that few people stopped to question, and that was bound by the planetary trilogy of Venus, Earth and Mars. Before anyone had time to express real doubt about the matter, textbooks had come to accept that, since the planets were still cooling off, those closest to the sun must be the hottest, and those farther away must be the coolest.

So they took another look at Venus, Earth and Mars. Each world had amosphere. Venus was the heaviest, Earth somewhere in the middle, and Mars had the thinnest layer of atmosphere. Venus was the hottest, Earth was temperate (to *us*), and Mars was the coldest. This *had to* be obvious, since the planet closest to the sun still received the greatest energy radiating out from our star.

So another speculative conclusion was drawn. By age, we could ignore the total time of existence of these worlds. We were really referring, said the authorities, to *aging*. Thus Mars was almost a dead world with its cruelly thin air and terrible temperatures, Earth was in the prime of its development life, and Venus belonged to an age that compared closely to what Earth must have been many millions and millions of years ago.

Before you could even think about it, Venus had become an "Old Earth" world, with a steaming atmosphere, dripping with water, covered with swamps, rocked with volcanic eruptions, and echoing to the thundering bellows of monstrous creatures not too unlike (conveniently) the dinosaurs that once did their bellowing on Earth. Venus was what we once were and Mars was what we could become. And that was that, and don't bother us with other considerations such as the retention of an atmosphere being in direct proportion to the gravitational attraction of a planetary

PLANETFALL

body. Unburdened by any way to *prove* their ghastly swamp creatures of a terrestrial Paleozoic era, writers, drawing their facts from what little scientific information was available, launched themselves on a field day of exciting adventures with finned and scaled swamp girls fleeing local monsters and visiting astronauts from Earth.

Then everything came to a dead stop. In 1962 a windmill-shaped robot fell past Venus. Mariner II scanned the mysterious world with a small but vital battery of instruments and sent back the results of its studies. First, there seemed to be a thickness of clouds that defied all predictions, for no one had ever expected a cloud layer so dense that it would *start* at a height of forty-five miles and would be seventeen miles thick, reaching to a height of sixty-two miles above the Venusian surface (above Earth, such a height is equal to 100 kilometers, and is accepted as "outer space").

But what really shocked everyone was that Mariner II indicated a surface temperature of 800 degrees F. above zero. Impossible! That's what the water-world adherents insisted, anyway. Because if the temperature was that high, water couldn't exist on Venus, and at least thirty terrestrial elements, including zinc, would melt under those conditions.

The scientists controlling the Mariner II mission pointed to the instrument findings and added a few details. The surface (*if* Mariner II instruments were accurate, that is) was likely partially molten in many areas. Sand and dust made up the surface characteristics. Worse, the atmospheric pressure at the surface was at least twenty times greater than found at sea level on Earth, which meant a pressure of some 294 pounds per square inch.

Scientists, who had long before rejected the heavy water content of Venus, had anticipated, after careful studies of that world, that the atmosphere would be mainly carbon dioxide. The clouds, they believed, would also be the same, so they were caught completely off balance when the instruments showed that the main elements of the high clouds are condensed hydrocarbons, rather similar to smog.

But it was the temperature of Venus, and the insistence that if there was any water or water vapor in the atmosphere of Venus it could be no more than a thousandth of that on Earth, which brought forth new violent controversy. Many scientists said that Mariner II, scanning Venus from a hurtling rush through space, could not possibly be right in its findings, and they were going to stick to their beliefs that Venus was a world rich in water in both liquid and vapor forms.

(One finding by Mariner II was that Venus does not rotate with a speed different from its revolution—which, of course, was later proved to be completely in error. At the time of the Mariner II flight, radar bounce experiments off the Venusian surface also showed that rotation and revolution were equal.)

The scientists who refused to accept the studies of Mariner II went to work with a vengeance to prove they were right, and, if nothing else, they wrote one of the more interesting chapters in the studies of our sister planet. A scientific team from Johns Hopkins University sent special telescopes and instruments beneath towering balloons to a height of sixteen miles above the Earth's surface, so that they would be above most of the atmosphere when they were trained on Venus.

What they discovered—as determined by their instruments and their subsequent conclusions—emphasizes just how *wrong* the most careful scientific investigation can be. Or, if you're expecting to find something you really *want* to find, you can make your instruments prove almost anything.

After the flights in 1964 of the huge balloon-borne instrument packs, the team of scientists, elated at their findings, announced that Mariner II was absolutely wrong about water on Venus. The telescopes suspended from the balloon had showed that the spectra reflected by the Venusian cloud cover were almost exactly the same as the spectra of ice crystals on Earth.

PLANETFALL

Which meant that not only Mariner II but all radiotelescope observations of Venus were terribly wrong. Venus, proclaimed physicist John Strong, was blanketed by clouds saturated with water vapor. They were like those of Earth. They were *not* poisonous gases and dust, no matter what Mariner II reported.

Dr. Strong, who was the director of the Johns Hopkins Astrophysics and Physical Meteorology Laboratory, stated: "Since it is known that carbon dioxide exists on Venus, proof of water vapor forces us to re-examine every previous calculation made concerning the possibility of some sort of life existing on the planet."

Based on that sort of statement, scientists could be expected to storm the walls of government to obtain huge grants that would let them send one robot after the other to Venus. Not simply to sail past the planet, but to slide into orbit and to descend into that maddeningly opaque atmosphere. It would seem, with water so abundant, that the greatest chances of life in our solar system elsewhere than on Earth must be on Venus.

Yet Venus was kept to a priority somewhat lower than Mars, which got the lion's share of appropriations for robot studies. The arguments raged to a high pitch; those who ridiculed the balloon-telescope studies pointed out that even if there was heavy water vapor in the clouds, Mariner II had indicated temperatures of 800 degrees F. Life as we know it can't even get going, let alone develop, under such heat. The stable chemical bonds in the giant molecules that characterize Earth life could never form.

The Johns Hopkins group brushed aside this problem by the simple procedure of saying that the Mariner II temperature readings were wrong. Space probes and ground-based radio telescopes measure the temperature of distant objects, such as Venus, by studying the electromagnetic radiations emitted by the planet. These are actually thermal radiations that lie within the spectrum being studied, and Mariner II and the radio telescopes, while not matching each other exactly, left little question but that Venus was extraordinarily hot. Too hot for Earth-type life forms. Period.

SISTER WORLD

The Hopkins scientists retorted that if the Venusian clouds contained water, then such clouds can emit radiations that don't necessarily have a thing to do with heat. They emphasized that lightning is an excellent example of a misleading emitter. What seems to have happened, claimed the Hopkins team, was that the nonthermal radiations of Venus had misled Mariner II and the radio telescopes. "Such radiations," explained Hopkins team member William Plummer, "could make Venus look twice as hot as it really is."

It was an interesting time for all concerned, but in the end, the Hopkins team went down without a murmur. The argument waxed hotter and hotter, until just about everyone agreed that the only way to prove once and for all the temperature and atmospheric composition of Venus was to, first, send a package of instruments into that atmosphere, and, second, land a package of instruments on the surface of Venus.

Not exactly an easy task. But the Russians did it much sooner than anyone expected.

CHAPTER 11

Into the Atmospheric Deeps

Let us—because the choice is ours—jump ahead once more in time to the flight of the *L. Gordon Cooper* as the great nuclear-drive spaceship nears the second planet from the sun.

We'll be back to the elaborate robots that were the first to plumb the thick clouds of Venus. But first we can enjoy the opportunity to travel far ahead in time and join a manned expedition to that planet. An expedition unlike any other ever risked by men.

We're catching up with Venus. The mechanics of our space rendezvous bring the nuclear spaceship in a chase mode. Moving with greater speed than Venus in her orbit about the sun, we catch up steadily as we fall inward toward the sun. From the observation ports, Venus looms larger and larger, a great half-world, one-half toward the sun a blinding yellow-white, the other half shrouded in darkness. It's like seeing a smooth-surfaced half-moon that suddenly glows with a hundred times more brilliance than we've seen on Earth's satellite. You can't help making such comparisons. We always need something to which we can relate when we see the unexpected or the startling, and Venus, if nothing is, *is* another world, both fascinating and extremely dangerous.

In the control room the computers dictate everything. Radar beams flash the distance from Venus, the digital numbers changing steadily. Distance, angle of closure, rate of closure, time to

INTO THE ATMOSPHERIC DEEPS

ignition—it's all there. The FDAI gets special attention, of course, but the men at this point are simply observers, and they won't interfere with the automatics running the ship just so long as everything continues to function so well. It's been a long time since the first space computer systems, and accuracy and reliability have reached levels that were only dreams back in the old Apollo days.

Even though the nuclear drive will come on with low thrust at first, increasing steadily as we near Venus, it's been quite some time since we've felt any acceleration force, and everyone is under orders to strap in. You've seen it before, but there's still that fascination when the drive comes on and you stare at the view screens, watching the televised plume of thin violet fire spuming away from the nuclear chambers. Not behind you, for *L. Gordon Cooper* is decelerating. The trick now is to reduce the sunward velocity, when very close to Venus, so that the big spaceship drops neatly into its celestial slot to take up orbit around Venus, just as the Apollo spaceships did around the moon. Thirty-eight long minutes later it's done. We're in orbit at a height of 164 miles above the surface.

It's an incredible sight. You've seen Earth from orbit in all its glory, and you've swung through airlessness around the naked, battered moon, but beneath you now is a world almost as huge as Earth and completely hidden from all sight by the thick and massive cloud layer. The spaceship falls in its breathless swoop at nearly 17,000 miles an hour from the daylit side of Venus into nightfall. Just about forty minutes later you're waiting for *the* sight, and it comes with a breath-taking shock. The Venusian dawn, as seen from orbit, exceeds anything ever known while looking down upon Earth. For the sun, still far from sight, seems to set aflame an atmosphere that rises more than three hundred thousand feet above the world below. The thick cloud layer begins to gleam a dull silver, then, almost swifter than the eye can follow, the silver slides into an utterly overwhelming red. You remember that the sun is only sixty-seven million miles away and the effects are greatly intensified. The red brightens to orange and then to the dominant coloration of the Venusian clouds, a yellow that is pure gold and bronze and colors beyond all description, and it's more than this mantle of burning gold covering the world, for above the heaviest mass of atmosphere are the swirling gases always in swift motion around Venus, and they're *alive,* twisting and spinning impossible tendrils of glowing light that you can see only for minutes before your speed and the sun exploding upward change everything. You've forgotten to put on dark glasses, but this ship thinks for you. Just before that moment when the sun could become harmful to your eyes, the viewport polarizes, and you're looking through suddenly darkened glass that makes the sun a coppery smoked disk. You can hardly wait for the sunset . . .

Someone came up with the name *Eagle* and it stuck. The name of the first manned spacecraft to descend to the surface of the moon in July of 1969. Now we're going to attempt the first landing on the planet Venus, and there seems to be an element of good fortune attached to resurrecting the name of the first Lunar Module to kick up moon dust.

But there isn't the faintest resemblance between the two. The first *Eagle* was a crazy spiderlike affair with long, crooked legs and curving landing pads, an angular body with all sorts of sharp edges, antennae sticking out in all directions, and a metal skin so thin (to say nothing of the gold foil crinkled about the ship for thermal protection) that a man's fist could have punched through. *Eagle II* is the opposite. It's almost impossible to say what it *is,* though. The first ship destined to take men to the Venusian surface is a crazy mixture of a spaceship with a powerful nuclear drive, a heavy mass that descends for part of its trip by parachute, and the manned assembly, which is the craziest of all.

It's a perfect sphere. You think of spaceships, and you think of Mercury, Gemini, Apollo, all of them through Skylab and

INTO THE ATMOSPHERIC DEEPS

Shuttle, and the idea of a sphere is crazy. But there's Vostok, remember? And Voskhod. They were spheres. The Soyuz ships were almost spheres; only the reentry heat shield marred the shape. But all that is coincidence.

Eagle II isn't patterned after a spaceship. It's copied, in terms of strength and capability, after the *Trieste II,* the bathyscaph that was the first deep-sea vessel to descend all the way to the bottom of the Pacific Ocean in the Challenger Deep. Where *Trieste II* went, the pressure was sixteen thousand pounds per square inch.

It won't be that bad for *Eagle II,* but fifteen hundred pounds per square inch is nothing to ignore—and that's the pressure along the surface of Venus. A hundred atmospheres as we know atmospheric pressure at sea level on Earth. But the ride of *Eagle II* is infinitely beyond anything known by the bathyscaph that plumbed Earth's oceans. There the ride was gentle, carefully controlled.

Eagle II will be, compared to anything ever known on the terrestrial oceans, or even within them, a nightmare. As the man who plays a combination of roles—handling cameras that take pictures and film in visible light, in infrared, in radar and sonar scanning—you'll probably be able to see and watch more of what's going on than any other member of the six-man crew. They'll be busy handling matters that could be spelled out as life or death. You'll record what's happening and hope it won't become the final chapter in the epitaph of all concerned . . .

It begins. *Eagle II* is nudged gently away from the huge nuclear mother ship. You drift alongside while every last item is checked and rechecked, because there's no margin for error where you're going. Everything is in order, everyone is strapped in—*tightly*.

The retrorockets fire. Not the nuclear drive of your spacecraft, but solid rockets that hurl flame for thirty seconds and then die out, to be jettisoned as so much garbage. They've done their job. The orbital balance is a losing game to Venus's gravity, and

Eagle II is on her way down into that impossible boiling soup of an atmosphere. For quite a while you're going to be blind, in the visible light spectrum. But with radar, infrared, pumped lasers, sonar and other instruments, you'll all know when the Venusian mountains rise from the swirling dark mists so that (you hope) they can be avoided.

The ship is turned to take the fierce, long and fiery blow of entry into the atmosphere across the disposable shield that will precede *Eagle II*. Like the others, you attend to those instruments for which you're responsible. But your eyes keep glancing at the master dials. There it is: the .05g light just came on. One-twentieth of a gravity. The same signal that marked deceleration for the very first manned ships so long ago. It's going to be a beaut, with that pea soup of an atmosphere out there. It comes quickly. The acceleration—it feels the same as firing upward—crashes into and through you. The semisupine couch takes much of the pressure, but it mashes you into the flexicushion and it's *rough*. Through the ports you see the orange flame as the thick atmosphere of Venus is taken by the ablative heat shield and transformed into fiery hell. Inside there's a dull thundering roar and *Eagle II* trembles through every metallic sinew. Then the worst of it is over. You brace yourself as the pyrotechnics fire and the blackened heat shield is blown away to the side.

Everything happens with efficiency and exact timing. You hear and feel the *thumps* of the reaction control system firing automatically, responding to the commands of the computer and the FDAI, which has been slaved to the true horizon of Venus, to keep *Eagle II* balanced precisely along her constantly shifting flight course.

The drogue chutes go out. Another hard thump inside the ship's structure as an explosive slug fires away the long, small drogue chutes that stabilize and slightly decelerate the massive spacecraft. The temperature's going up. You're through the cold part of the clouds and dropping rapidly into the hissing, broiling atmosphere of Venus. And this is the biggest difference between

INTO THE ATMOSPHERIC DEEPS

the old bathyscaphs and what you're doing. *Eagle II,* among other things, is a huge freezer working under maximum output. The nuclear power plant is needed for a lot more than space drive. It runs the cryogenic systems and the electromagnetic buffer field to protect the spacecraft against temperatures under which lead and zinc run molten. You took more heat in penetrating the atmosphere, but that was ablated away, and it didn't last too long. When you get into this atmosphere the problem is worse, because you *stay* in the heat, and the ship begins a process of heat-sink, and the temperature of everything starts going up. Unless those cooling systems and the electromagnetic field radiate away your own heat—the internal heat of the ship as well as that from the atmosphere—you'll all become part of some giant stew. And you've got to protect not only the pressurized crew compartment but the instruments, the nuclear drive section, the engine chambers, fuel storage, pressure and electrical lines—the works. While all this is going on, you keep in mind that the outside pressure is increasing. That isn't vacuum, what every good spaceship likes to have about it.

It is cruel, relentless pressure that could crack like an eggshell any weak point in *Eagle II.* Man, you're not a spaceship any more—*you're an interplanetary submarine*!

Almost, but not at this moment while you're still falling through the Venusian atmosphere that seems to thicken perceptibly every thousand feet on the way down. The recording cameras get a good picture of the drogue chutes, three small ribboned canopies fluttering tautly high above at the end of their lines. It's already getting darker and darker, and floodlights playing against the canopies bring them out of increasing gloom. They look like three fishbelly-white eyes glaring down on the spacecraft.

"Drogues away; main chute coming out reefed." The words first, calling off the check list item. Another dull sound thumping through the ship and through the port; sighting along with the cameras you see the drogues suddenly whip away, high and to the side (a great shot, that one). Moments later the main chute booms upward, reefed tightly to keep down the sudden impact of opening. *Eagle II* sways suddenly, then settles down. Hang on, you say to yourself; the explosive squibs cut the reef lines and the great ribboned chute is out.

Now you're falling with more control in terms of descent, and you hope the plastimetal material of the chute can withstand the temperature rise, because it's already 390 degrees F. out there and going up steadily.

Eagle II rocks sharply. You watch the instruments and listen to the crewmen and it—is—not—good.

The radar altimeter beaming downward isn't working well at all. A combination of temperature and pressure was expected. But not the horizontal winds! The best that can be determined is that your ship is pounding over the surface with a speed of more than 165 miles an hour, and buffeted by violent thermals. The downward-scanning radar is being messed up by thermoclines, strange levels of heat and other levels of colder air. The result is that the atmosphere is wildly stratified in terms of pressure and temperature, and the winds are carrying before them strange gases and an unknown quantity of dust, or sand, or *what* . . . The result is you're simply not sure of how high you are above the surface, and there are mountains down there, great big rocks that may stand anywhere from four to twelve miles above the surface.

The outside pressure gauge shows you're pretty far down. The reading indicates two items. The first, in terms of atmospheres, shows 802, which means that *if* the probes that preceded this mission were accurate, then you're getting close to the median surface of Venus, but you may already be *below* the higher peaks. And if you slam into one of those at this horizontal speed, well, that's *it*. What about that second gauge? It's been changing steadily and it's already indicating its digital read-out of 1149. But it may be lagging, and the pressure may be higher than 1,149 pounds per square inch.

The decision is made to get away from the parachute and go

to onboard drive. Everyone is relieved at this, for more than safety is involved. Survival is at stake.

The nuclear drive has been active all this time, of course, and now it is brought to full power. But there are several nuclear drive chambers, and one of them is critical this deep within the atmosphere. It operates as if *Eagle II* were a submarine—which, at this moment, it certainly is.

Outside the thick, supercooled hull are the equivalent of hydrojets. They're a form of jet engine adapted for deep water propulsion in Earth's oceans. They move water through a jet chamber, and get full control, including hovering ability, from the reaction of water through the jet chambers. The submarine using the jets, of course, contributes its own buoyancy.

The deep, powerful whine of the atomic hydrojets outside *Eagle II* penetrates right through the hull, but it's a welcome sound. You stare at the instrument dials. Ground speed is slackening steadily, for the ship is now heading into the wind, and the nuclear hydrojets are reducing backing speed. It's like an airplane flying sixty miles a hour into a hundred-mile-an-hour wind—your air speed is sixty miles an hour, but you're drifting *backward* over the ground at forty miles an hour. The power increases until the backward drift of *Eagle II* ends. Now forward speed picks up with more power. Ground speed where you want it— straight ahead—is twenty miles an hour. A safe speed, slow enough to stop or veer away from anything that might emerge from the soup-thick mists, and——

Suddenly you realize what's happened. At this moment you're in a submarine. The ground—if it can be called that—is barely a hundred feet below. And the true impact of your environment sinks in.

The pressure outside your hull is the same as that found 2,600 feet beneath the ocean surface on Earth! That's a pressure great enough to smash in the thick hull of a large nuclear submarine. Only deep-diving research subs, designed especially for that sort of cruel pressure, could survive. Outside your hull there's atmosphere, but it's beyond your ability to conceive of *air* at 1,544 pounds per square inch!

And the temperature . . . it's enough to scare the wits out of you. From the pressure readings it's obvious that *Eagle II* has reached the lowest part of the surface, where the pressure is greatest. The gauges indicate a temperature of 1,071 degrees F., and in your mind you try to imagine lead and zinc and other metals first flowing as molten liquid and then starting to vaporize.

The atmosphere is as alien as the pressure and the temperature. Ninety-six percent carbon dioxide. The remainder is a smattering of inert oxygen and other gases, with no more than a wisp of oxygen. Barely a hint of water vapor. Plus poisonous metals boiling off and volcanic gases spewing into the massive soup of atmosphere. This world, once conceived by so many as a watery Earth of eons past, has turned out to be—literally—a perfect rendition of Hades.

Now *Eagle II* is perfectly into its exploration mode. You're about one hundred feet above the surface, but seeing is almost impossible. No thick and obscuring clouds about you; those are many miles overhead. But in the optical sense it's like being at the bottom of a half-mile of water. Most of the sunlight reaching Venus is reflected away by the thick cloud cover, and what's left is drained and thinned out by the time it gets to the surface. The atmosphere is so thick that its molecules, to say nothing of gases, dust, sand and debris, manage to scatter light until it's no longer light as we know it. Ever drive through a thick fog and turn on your bright headlights? Remember what happens? The light is reflected so wildly you can hardly see anything. You go to low beams to reduce the scattering. That's your problem now. *Eagle II* has batteries of floodlights for the cameras, but they're not working very effectively.

You're drifting forward slowly, upheld by the thickness of atmosphere, the buoyancy of the spacecraft/submarine and the nuclear-powered hydrojets. Perfect for seeing. But the ground is a hundred feet below you, so slant vision makes the distance even

INTO THE ATMOSPHERIC DEEPS

greater, as you try to see ahead of the ship. The floodlights are simply scattering light like silent swirling explosions. Well, it's not entirely unexpected.

Down goes *Eagle II,* slowly and with great care. Laser scanners and sonar establish a safe probing of any unseen obstacles. As you get closer and closer to the surface you finally begin to make out some details. You wonder if anything could ever evolve in this hellish, boiling, poisonous soup. Now you can see the ground directly below. *Eagle II* moves forward with very slow speed so you won't blur details with movement. The still cameras are snapping away in ultraviolet, infrared, visible, radar and other frequencies, and you hope they're getting it all down. Later you can piece together what the camera batteries are recording, and come up with an accurate portrait of this astounding world. But now you want to *see.*

You order special flares released from their dump tubes along the bottom of the ship. Others are fired with slow-burning solid rockets to move ahead and above the ship, and they drift back toward you. With the intense light of the flares at some distance, coming back, there should be enough light to get some visual acuity for your eyes and the cameras. Out go the flares, and the murky gloom of Venus yields to the intense light scattering in all directions. You're also using the floodlights directly ahead and downward.

You *see* Venus for the first time. Or, at least this one tiny part of Venus. A strange sort of sandy dust. Rocks of odd shapes, most of them buried. Dunelike swirls. Movement over there to the right. Quick . . . turn that way!

Movement? That's impossible, it can't be . . . The ship glides in that direction, but the glide is changing to a bumpy, rocking passage. And there's noise, strange growls of thunder. Turbulence suddenly. There. You get much closer than you should. Bubbling pools, throwing up thick and viscous—what? You can't tell. It could be steaming, boiling mud, or liquid metal, or a mixture of mud and lava and metal and sand and, well, you end that sort of speculation. The instruments taking in every detail will spell it out later.

So far you haven't seen anything except what was directly in front of your nose, so to speak. What you want most of all is a fairly good picture of what lies well ahead of the ship. A panorama. Out go the flares again and—suddenly you've got the best visibility for which you dared to hope. You're stunned by the sight, because you still can't see!

The atmosphere is so thick that the light all about you bends sharply. It's twisted and distorted. Disappointed, frustrated that you can't see the horizon, you know that even if the horizon were visible you still couldn't see it. One of the strange paradoxes on this strange world. If it were possible to see across a distance to the horizon (there, that's better), you wouldn't see the horizon itself, except as crazy, wavy lines sweeping sharply upward.

The only way you or anyone else is going to "see" the surface of Venus is by going over all the camera and instrument data after you're back on Earth, and creating a picture—an artificial picture, in effect—that you can see in visible light.

Well, at least the instruments are taking everything down. You start to relax, concentrating on watching the equipment, when suddenly the spaceship/submarine rocks sharply. Deep thunder crashes all about *Eagle II,* thrumming through the structure. Outside: intense flashes of light. No way to know if it's static discharges of a scale beyond anything known on Earth. It could be volcanic or even, more likely, a combination of the two. Whatever it is the shock waves are enormous, and a glance at the environmental instrument panel tells everyone what's happening. The shock waves are reaching the maximum intensity point. If this keeps up you won't be able to count on the systems of *Eagle II* to survive the hammer blows reaching out to envelop the ship.

Time to execute the well-known maneuver known as "Let's get the hell out of here." Which is something of a joke, because it ought to be written down as "Let's get out of Hell," which

INTO THE ATMOSPHERIC DEEPS

would be more appropriate. The hydrojets whine fiercely under increased power. It can't be too much, because of that thick atmospheric soup and the crashing shock waves. But up you go, and with every minute the pressure eases off as you gain altitude. Miles above Venus you're back in that thick belt of clouds, and the hydrojets are losing their ability to keep the ship climbing. Now it's time to convert. You've been a submarine in mercury-thick atmosphere. Now you're high enough where you've got to become spaceship/submarine, and the main nuclear chambers crash into life. You've never heard them like this, but sound carries powerfully through the soup outside the hull. Carefully, very carefully because of the still heavy pressure, *Eagle II* rides her nuclear thrust higher and higher.

The computer is handling the dangerous and sensitive job of transitioning from atmosphere to vacuum. Visually, you're still in clouds more than two hundred thousand feet high, but the pressure is low enough now to really pour on the juice. The hydrojets are silent, useless this high. The computer's instruments sense the swiftly decreasing pressure and the time for full thrust. Violet flame slashes back through the remaining thin clouds, and *Eagle II,* as if she were grateful to be free of that awesome pressure below, slices higher and higher, arcing over to match the horizon far below, and accelerating steadily.

There it is: orbital velocity. The drive shuts down and you're in free fall. Weightless. Everyone sighs with relief. It's almost as if you've escaped the planet below. Now the computer will work out the intricate details of rendezvous and you'll dock with *L. Gordon Cooper.*

Venus?

That's a world for robots to explore. Just like the one in which you're safely cocooned.

They can have Venus. Sister planet or not, it's no place for men.

Not with the jewel we call Earth as our home.

CHAPTER 12

Those Incredible Robots

Old sayings stay with us most of our lives, and one that I try to keep close to mind is that familiarity breeds contempt. The more we're exposed to an event, no matter how stupendous or overwhelming it may be in its introduction, extended repetition of that event, the familiarity of a machine, or an object, renders it prosaic. It takes a deliberate effort, sometimes, to drag yourself back to wonders that have been experienced in your past.

So, every few months I drive down the eastern coast of Florida, along Highway A1A, into Patrick Air Force Base. This is the headquarters of the Air Force Eastern Test Range, with Cape Kennedy (today it's again Cape Canaveral) eighteen miles along a curving shoreline to the north. Parked along Patrick's flight line are huge bulbous-nosed jet aircraft, Boeing C-135 giants modified to act as ARIA (Apollo Range Instrumentation Aircraft), communications links during the manned flights to the moon, and, during those times that men thunder between Earth and heaven for the Skylab space station program.

My destination within Patrick is a building I saw erected many years before we ever had a moon program. The Technical Laboratory, looming over the beachfront, and still host to many classified and well-guarded military space activities. But my interest today lies not in what goes on within that building but in front of the massive structure.

My interest is history. Standing in front of the laboratory is a row of missiles and rockets. Starting at the south end is a winged

THOSE INCREDIBLE ROBOTS

Bomarc, then one of the early Polaris missiles, and then a Pershing battlefield missile. There's also a winged Mace and the first intercontinental winged missile that went through grueling birth pains, the Snark. Moving north we see a Thor. The missile in its original form, with a thick and heavy nose cone at its top. It's time to stop a moment, for this missile, after it had been modified and additional stages were placed atop the assembly, became not a bigger missile but the first stage of a rocket that would become known as a workhorse of space. It seems almost to give off an aura of history. The Thor-Able . . . first to reach out to the moon. Then the Thor-Agena for our Discoverer program and the first satellites to be brought back from space. Thor-Ablestar, Thorad and Delta. One payload after the other into Earth orbit, scientific probes to the moon, more probes into orbit about the moon. Weather, communications, navigation, solar, geophysical—and many other—satellites.

After Thor, a giant—a Titan I. The same rocket I saw launched from a distance of just over a thousand feet down a narrow road. The rocket that created *Aurora Titanalis* I described in chapter 6. Beyond Titan, the diminutive Minuteman. For the reasons I came here it seems out of place, knowing how many hundreds are buried in massive silos deep beneath the ground, the warheads crammed with hydrogen bombs.

But directly beyond Minuteman *is* the reason for coming to this place. A rocket that once, long before Saturn V, truly seemed a great giant. Now it no longer holds that distinction, has not, really, for a long time. Other rockets have surpassed it in size and thrust and capability. But if it no longer is a giant in those characteristics, it is a giant in history, and the road away from Earth is straddled by this epochal assembly of thin aluminum skin and three flaring engine chambers.

This is the Atlas-Agena, and it *is* history. It changed the horizons of man and opened up to him staggering knowledge. It carved a path to the moon and stripped away the mists of time and distance surrounding Venus and Mars.

PLANETFALL

The first-stage booster, the Atlas, was the first Air Force satellite to go into orbit. It carried our Mercury astronauts into their fiery rides at five miles a second. With upper stages it sent Ranger to the Moon, and Mariner to Venus and Mars. The combination of Atlas and Agena was the key.

No more Atlas-Agena these days. A museum piece. But Atlas lives on as the first stage for Atlas-Centaur. Heavier payloads with this combination. Surveyor to the moon, bigger and heavier Mariners to Venus and Mars and Jupiter and Saturn.

We aimed Mariner I for Venus. High above Earth's dense atmosphere the guidance system failed. The rocket veered from the precise course on which it had been sent. On Cape Canaveral a man sealed within a control center known as the Green Room felt a wave of agony and then did what he must do. He opened a safety cover and stabbed down on a button, and a dream vanished in a mushrooming ball of dark flame. The date was July 22, 1962.

Far from Cape Canaveral, in Moscow, Russian scientists received word of the American failure. They understood all too well the feelings of the American scientists involved. For the Russians had known bitter failures of their own—too many of them. But the years have shown an extraordinary determination on the part of the Soviet Union to bridge the awesome gap between worlds.

For a full *decade* Mars eluded the heavy robots of the Russians. On October 10 and 14 in 1960, their first shots to the red planet exploded or otherwise failed during the launch phase.

Two years later, on October 4, 1962, another shot to Mars failed during launch. A week later, on November 1, they sent their first gleaming robot racing toward the fourth planet from the sun. For nearly five months the robot sent back data on conditions in space, and then went silent as the communications system failed. The silent spacecraft sailed past Mars at a distance of 120,000 miles.

THOSE INCREDIBLE ROBOTS

The Russians were making an attempt to send grouped ships to Mars. Three days after their first successful planetary launch outward from Earth, they fired yet another Mars probe, on November 4, 1962. It failed during launch.

On November 30, 1964, the Zond 2 probe roared away from Earth. A successful launching toward Mars. Months later the communications systems of Zond 2 fell inexplicably silent, and the brilliant robot, now beautifully shaped but useless metal, disappeared into its solar orbit.

Success would continue to elude the Russian attempt to tear away the veil from Mars until May 19, 1971, when a five-ton probe called Mars 2 cracked upward from its launch pad. A little more than a week later, on May 28, Mars 3 made its two ripple-fire successes as it rifled away from Earth. Both probes would go into Martian orbit. Both probes would send small ships from that orbit into the Martian atmosphere, to descend to the surface.

What the Russians tried to do with Mars, with one failure after another, they attempted as well with Venus. Here they would come to know both bitter failure and extravagant success, and they would be able to write into their historical tomes that Soviet spacecraft were the first to reach the surface, not only of Earth's moon, but also of Venus and Mars.

The first Russian probe to Venus was launched on February 4, 1961, and failed as it struggled to break free of Earth's massive gravity. Eight days later a second shot was made, and as it set sail on its course for Venus, after a beautiful launch, it was named Venus 1. Fifteen days later the power source aboard the heavy robot went dead. The mute spacecraft would pass in complete silence beyond Venus at a distance of 62,000 miles.

On August 25, 1962, the third shot failed during launch. A week later, on September 1, another failure. And yet another, twelve days later. The Russians tried again for Venus on April 2, 1964, with a probe called Zond 1. It made the perilous upward journey from Earth with success, and then, the same failure that seemed to curse the Russians—the communications system broke down.

On November 12, 1965, Venus 2 was launched. Again success was in the grasp of the Russians and again it was lost. Just before reaching the vicinity of Venus, with great accuracy in a path that would take it within 15,000 miles of the planet, all communications were lost. There had been another launch on November 16, 1965—it failed in a way guaranteed to test a man's faith. On March 1, 1966, the probe named Venus 3 tore into the atmosphere of Venus in a dazzling performance of long-range accuracy. Both Venus 2 and 3 during the long flights had buoyed Russian hopes with excellent performance, sending back a steady stream of data on interplanetary fields of magnetism, cosmic rays, radiation from the sun, micrometeoroids and other areas. Venus 2 and 3 represented yet another satisfaction for the Russians; their attempt to gather information on Venus with these two probes had been followed by a third, launched only a week after Venus 3 was on its way to that planet. The November 23, 1965, launch failed on the way out from Earth.

But here were the other two spacecraft, flying precisely where they had been aimed, cameras and other instruments at the ready. And just before encounter with the sister planet of Earth, all power systems or communications systems inexplicably went dead, and nothing was heard from far off in space.

There would be more launch failures, but against the success the Russians grabbed suddenly after many years of frustration, these would be forgotten. On August 27, 1962, the United States sent Mariner II on the way to Venus, and on December 14 of the same year, the space probe passed within 21,645 miles of the planet's surface, sending back to Earth, aside from other data, the electrifying news that the surface temperature on Venus averages 800 degrees F.

What Mariner II reported was noted carefully by the Russian scientists, for it enabled them to alter their own instrumentation. Knowing generally what to look for, they could more carefully and effectively direct their research program with their own probes. They noted the findings of Mariner II. During the 109-day interplanetary flight, the 447-pound spacecraft had detected, and relayed to Earth, that a faint interplanetary magnetic field persists along the ecliptic plane; that a constant solar wind flows from the sun with speeds up to 480 miles per second; that the quantity of cosmic dust (micrometeoroids) is much less near Venus than it is near Earth; and that the intensity of cosmic rays, the powerful particles that originate from somewhere beyond our solar system, is nearly constant.

Then Mariner II fell past Venus and sent back its report on the planet. Venus is blanketed with cold, dense clouds in the upper atmosphere; the surface temperature is 800 degrees F.; the temperatures are essentially the same on the dark and sunlighted sides of the planet; a cold spot was found in the cloud tips; there was no measurable magnetic field, nor was there a measurable radiation belt—the latter two items characterizing the physical environment of Earth.

The Russians had reached Venus with exact precision in their probes, and now they were ready to put their failures behind them. On June 12, 1967, they launched Venus 4 (also called Venera 4) on a 217,000,000-mile looping flight to the distant planet. Two days later the United States launched Mariner V to Venus; it would arrive just thirty-seven hours after the Russian probe punched into the atmosphere.

For Venus 4 was a towering success. Its findings were staggering, inasmuch as they provided the first *direct* scientific studies of the planet by penetrating and then descending through the atmosphere, transmitting data from a whole battery of instruments during that descent. (Mariner V passed within 2,480 miles of Venus, carrying out the same manner of investigation as its Mariner II predecessor, but with finer detail.)

Venus 4 is also a prime example of how each new mission builds on the success and failure of previous flights. The 2,433-

PLANETFALL

pound spacecraft reached Venus exactly on schedule and then ejected an instrument package. As it slammed into the thick atmosphere the instrument unit was squeezed with a deceleration force of *several hundred times* the gravity force on the Earth's surface. After deceleration, a large parachute opened. Pocket-sized radios transmitted instrument findings steadily in the first scientific broadcast from another planet to Earth.

Until then any measurement of the density of the Venusian atmosphere was an educated guess. It was estimated as anywhere from the same as sea level on Earth to one hundred times that value. Very quickly, after the data was received, it became clear that the atmosphere was extremely dense. Venus 4 transmitted for ninety-three minutes, and ended its transmissions well before the surface was reached. Which, in itself, made it clear that the descent was extremely slow, testifying to the atmospheric density.

Many scientists believed that Venus had an atmosphere with no more than ten percent composition of carbon dioxide. Venus 4 attested to approximately ninety percent carbon dioxide as the main atmospheric constituent. Nitrogen, water vapor and oxygen were present only in very small percentages. The temperature values seemed strongly to confirm what Mariner II had reported long before—too hot for any life as we know it. And as to pressure, Venus 4 ended its transmissions just at the time the instruments transmitted a reading equal to eighteen times the pressure of sea level on Earth. The highest temperatures at this point were 536 degrees F. It was clear the descending instrument package was still high above the surface when it stopped transmitting, so that the values for temperature and pressure were almost certain to be higher, closer to the surface. There was one strange discrepancy between the findings of Venus 4 and Mariner V. The Russian probe detected a slight increase in a magnetic field as it approached Venus, but found no magnetic field close to the planet or within the atmosphere.

Mariner II had shown no magnetic field that could be measured when it swept past Venus in 1962 at 21,645 miles. But

THOSE INCREDIBLE ROBOTS

Mariner V, passing by at a distance of 2,480 miles, seemed to indicate a definite magnetic field. The question could be resolved only in later missions.

But it would be Russian instruments, for Mariner V was the last American probe sent to Venus to that date. (Now the next launch inward to the sun, a Venus/Mercury mission, is nearing its final preparations. An Atlas-Centaur is scheduled to fire the Mariner Venus/Mercury spacecraft from the Cape on November 3, 1973—at a time when the Russians also will be sending heavy robot spacecraft to the same planetary target.)

On January 5, 1969, the sky over the Tyuratam-Baikonur space center in the USSR rocked with the thunder of Venus 5 on its way to the cloud-shrouded world. Five days later, thunder crashed downward again as Venus 6 was sent on its long path, to follow close behind its predecessor. The failures of the old days were now, happily, a memory, as the two heavy spacecraft sailed with unerring precision to their distant objective closer to the sun. Each probe weighed 2,486 pounds, about fifty-three pounds more than Venus 4. And, like the first successful flight, Venus 5 and 6 were intended to transmit as long as possible during atmospheric descent. Venus 4 had indicated a pressure of eighteen atmospheres when it went silent; the two new spacecraft were much stronger so that they might withstand much higher pressures. Venus 5 reached the planet on May 16; Venus 6 arrived one day later. They both descended on the night side of the planet.

Each robot weighed—for the part that descended into the atmosphere—891 pounds. The oval-shaped spacecraft were 3.28 feet in diameter.

Entry procedures to penetrate the Venusian atmosphere began when the spacecraft were still far from the planet—23,000 miles for Venus 5 and 15,500 miles for Venus 6. Each probe ripped into the atmosphere at an angle of sixty-two to sixty-five degrees, with a speed of 36,670 feet per second. The deceleration force

PLANETFALL

was about twice that experienced by Venus 4—or 450 times surface gravity on Earth. When the capsules slowed to 690 feet per second, the main parachutes were released. The parachutes, based on the findings of Venus 4, were only one-third the size of the chute that had lowered the earlier spacecraft, so that descent would be faster.

Venus 4 had transmitted for 93 minutes before the instrument compartment was crushed by the pressure of eighteen atmospheres. Venus 5 transmitted for 53 minutes along a descent of 22.3 miles, and Venus 6 transmitted 51 minutes along a descent of 23.6 miles.

As they dropped toward the surface, the pressure increased steadily. When the pressure level reached twenty-seven atmospheres, the top of the instrument compartment of each robot collapsed, ending all transmissions. When this happened, the radar altimeter of Venus 5 showed the capsule was about 80,000 feet above the surface. Venus 6 ended transmissions while only about 35,000 feet high, indicating to the Russian scientists that Venus 6 apparently was descending over high mountains. The two spacecraft were about 185 miles apart when they began their parachute drops into the atmosphere.

Since the spacecraft were still extremely high when they reached pressures of twenty-seven atmospheres, the Russians revised upward their estimates of surface pressure to perhaps one hundred times that on Earth. If Venus 5 was descending over lowlands, the pressure (80,000 feet below) conceivably could have been as high as 140 atmospheres. (Keep in mind the altitude difference between the Challenger Deep and Mt. Everest on Earth—a distance in height of some thirteen miles, with an enormous difference in pressure of atmosphere possible through this distance.)

Venus 5, still high above the planet, recorded a temperature of 608 degrees F. Extrapolating to the surface, this reading would have indicated a surface temperature 80,000 feet below of 986 degrees F.

THOSE INCREDIBLE ROBOTS

Much finer detail was obtained of the atmosphere. The capsules showed that carbon dioxide was present in a ratio of 93 to 97 percent of the atmosphere (Venus 4 indicated 90 percent, and Mariner V indicated about 84 percent).

Nitrogen and inert gases were present up to two to five percent of the atmosphere. Oxygen was present to a maximum of 0.4 percent, and water vapor was barely detectable.

On August 17, 1970, the Russians made the first of two more launchings to Venus. Venus 7 raced away from the Earth on the start of a successful mission, but a second probe fired five days later, after reaching Earth orbit, misfired (and was redesignated Cosmos 359). Venus 7 went on to make another precise flight to the planet, descended all the way to the surface and transmitted data for twenty-three minutes before brutal temperatures destroyed the spacecraft.

Another double mission was set up for early 1972, and on March 27, Venus 8 was launched successfully. Four days later Venus 9 was sent into Earth orbit of 300 by 127 miles. The attempt to boost out of this orbit to Venus failed, and the spacecraft broke up into four objects (the launch was redesignated Cosmos 482 after the failure).

Against the success achieved by Venus 7 and 8 it didn't seem to matter much. Since Venus 8 is the last probe to have reached that world, the findings of the spacecraft bring us fully up to date.

Venus 8 weighed 2,065 pounds in its complete form, with a descent/landing capsule weighing 1,089 pounds. The capsule was built in the form of a sphere so weighted that when it touched the surface it would roll to a position to point its antenna upward. A second antenna was carried, and when Venus 8 landed, this antenna was ejected to land several feet away from the capsule, connected by a heavy electrical cable to the radio transmitter.

The landing was made on July 22, 1972, in the most successful mission of all spacecraft sent to Venus. For this flight the capsule spacecraft had been redesigned extensively, based on the findings

PLANETFALL

of previous missions. In addition to the double antenna, it was designed to get into the lower atmosphere quickly. The parachute was reefed to permit a rapid descent through the hottest portions of the high atmosphere, then would open fully to slow the descent just prior to landing.

Venus 8 flashed through atmosphere for its landing only 1,800 miles from where Venus 7 touched down. But the landing of Venus 8 demanded extraordinary control, since it was aimed at the narrow, crescent-shaped part of the planet in sunlight, the part visible from Earth. On the surface, Venus 8 transmitted for 13.3 minutes through the primary antenna, sending out information on light levels, temperature and pressure. At that point, transmitting switched to the second antenna for twenty minutes, with studies of the surface. A final thirty minutes of transmission again came through the main antenna.

It took one hour for Venus 8 to reach the surface after slicing into the thick atmosphere. At thirty miles above the surface the capsule was whipped along by winds of nearly 120 miles an hour. Thirty-two thousand feet up, the winds had dropped to about seven miles an hour. They did not increase as Venus 8 descended.

Temperature on the surface was measured at 900 degrees F. The pressure where Venus 8 landed remained at a steady 1,300 pounds per square inch.

The soil beneath the heavy capsule was loose, and Russian scientists believe it to be porous. The chemical composition of the surface materials showed that at least part of the surface is similar to that of Earth. The landing area contained radioactive potassium, thorium and radium in approximately the same ratio in which they appear in volcanic rock on our own world. This was the first clear indication of geologic background, and provided good reason to believe that Venus, just like Earth, the moon and Mars, was once sufficiently heated for the crust materials to soften and to flow in molten form. As the crust cooled, heavy elements dropped toward the core of the planet, while the

THOSE INCREDIBLE ROBOTS

lighter elements carried radioactive material with them to form the surface crust.

By now the Russians had atmospheric element findings from Venus 4, 5, 6, 7 and 8, and the conclusions they drew firmly for general composition was 97 percent carbon dioxide, about two percent nitrogen and less than 0.1 percent oxygen. A very low percentage of ammonia was also discovered in the higher atmosphere.

Now, with further American and Russian probes waiting in the wings, Venus will be forced to yield even more of its secrets. For there are plans to place satellites in orbit around the planet, as was done with Mars, and to send even stronger, more efficient capsules to the surface, there to broadcast not for a span of time measured in minutes but in days.

In the meantime, powerful radar beams are used to study the characteristics of Venus as well as other planets. Such devices have gained a steady improvement both in technical capability and efficiency in use, and that's precisely what happened with the Goldstone Tracking Station (NASA) in California's Mojave Desert. JPL scientists, using the 210-foot dish antenna, made a new study of Venus, and what they came up with has, frankly, startled the scientific world.

Beneath those thick and opaque (to visible light) clouds of Venus, in the area studied—about 910 miles long and equal in size to the state of Alaska—*the radar beams picked out huge craters along the surface of Venus.*

The craters showed up along the equatorial region of the planet in what was known to be a basically flat area. A dozen craters were mapped by radar imagery, the largest of which is 100 miles wide but only a quarter of a mile deep. Other craters extended from 20 to 65 miles in diameter—and such sizes are very close to similar type craters already discovered on the moon and on Mars.

PLANETFALL

The best quality in radar mapping of Venus, in 1970, had a resolution of about 20 miles. The latest study, by scientists of the Jet Propulsion Laboratory, dropped that down to only six miles. Thus the dozen odd craters, from 20 to 100 miles wide, are believed to lie in an area where many smaller craters also abound, but which lie beyond the "reach" of the radar dish antenna.

Dr. Richard A. Goldstein of JPL noted that this "area of Venus appears to be as crater-infested as the moon."

Dr. Carl A. Sagan explained that the "same sort of debris that makes holes on the moon should be bombarding Venus." He went on to say that what was mapped on Venus by radar suggests most strongly that the craters appear to be formed by meteoric impact rather than by volcanic eruption.

Which raises the immediate question. Can huge meteors penetrate the thick Venusian atmosphere without burning up? The answer would seem to be a definite no. Could the craters have been formed *before* the thick Venusian atmosphere was created? That's the signal for much muttering and argument.

Well, we're at an impasse. If the craters were formed by meteor impact, and the atmosphere is too thick to permit that sort of impact, then where is the source for the craters?

Maybe the meteoric impact theory is wrong and they're volcanic. *Maybe.*

Because there's another puzzle. The craters seem to have been seriously eroded. Okay so far. But Venus has no water, and all measurements of surface winds by the Russian Venera probes indicate a very mild wind along the surface.

"It's a great shock to me," stated planetologist Steven Saunders. "The crater discovery is telling me that the erosion rates are much less than we thought."

Well, it could be——

Whoa; stop right there. The first pictures of Mars from Mariner IV, and the second series of pictures from Mariners VI and VII, were drastically misinterpreted in many ways, and they led

THOSE INCREDIBLE ROBOTS

us along corridors that later proved to be the worst possible turns in studying that planet.

Let's wait for the next look at Venus—Mariner X—which, "if all goes well," will be winging it past Venus in March of 1974.

The history of trying to wrest secrets from Mars is considerably less complicated than it is for Venus. As we've seen, the Russians spent a decade enduring failures in their hopes of getting probes successfully to the red planet, and succeeded for the first time with their double launch of Mars 2 and 3 in 1971. The United States had tried a double shot in 1964. On November 5, Mariner III went successfully into orbit, and then went dead, with the failure of solar cells to deploy. Mariner IV, launched on November 28, was a spectacularly successful flight, passing within 6,118 miles of Mars in July of 1965. Twenty-one photographs sent back to Earth destroyed long-held and cherished theories, for on those historic pictures were the first signs that Mars was a heavily cratered planet.

Four years later, another double shot came off with thundering success. Mariners VI and VII reached Mars in July and August of 1969, passing within 2,200 miles of the planet's surface, and transmitting two hundred photographs back to Earth. Those pictures stripped away another set of veils beneath which Mars had so long remained a mystery planet, and revealed a terrain at which the pictures from Mariner IV had not even hinted. It seemed that every time we got another really good look at Mars, the theories we had created to replace the older, no longer tenable, had to be chucked out the nearest window.

Mariner VIII and IX were to be a huge stride forward. No simple planetary flybys this time. Both spacecraft would be dropped into complementary orbits about Mars. They would take thousands of photographs to create an elaborate picture map of the planet, as well as study Mars through batteries of scientific instruments.

On May 8, 1971, at 9:11 P.M. EDT, Mariner VIII lifted from its launch stand at Pad 36A on Cape Kennedy Air Force Station.

PLANETFALL

Thunder cracked its giant fire whip across the Cape, and it was a shot more beautiful than most. Swamp and brush fires had sent a heavy smoke pall drifting over the Cape, and the golden flame of the Atlas was especially enhanced by the smoke and haze that seemed to burn with its own deep but transparent fire. Sometimes what starts out with utter beauty and perfection is its own curse. Atlas burned all the way, Centaur separated and ignited, and I watched through binoculars the twin disks of Centaur burning. And then twenty seconds later the guidance system threw up and Centaur skidded into a tumble, with flame spattering uselessly in near vacuum.

Mariner IX was ready to go, but engineers exercised prudence to study carefully the cause of the failure (a small diode in the guidance package) before committing to the bird on the pad. In the meantime, as we waited, the Russians kicked off their heavy (more than five tons) Mars 2 and 3, each carrying a smaller ship to land on the Martian surface. Both shots went beautifully, and they would fulfill their promise.

May 30 was the day to commit with Mariner IX, and at 6:23 P.M. the Atlas-Centaur broke from its launch stand and cracked neatly into space, slinging the payload on its way for a 247-million-mile flight that would bring it to Mars just before the arrival of the Russian spacecraft.

The degree of accuracy our space engineers sometimes achieve with their complex missions is too exquisite really to comprehend. Mariner IX left Earth on May 30 and arrived at the red planet November 13. The small engine burned for exactly 915.6 seconds, and when flame ended, Mariner IX was in orbit about Mars, *only 4.4 seconds earlier* than had been designed into its flight plan so many long months before!

After it was locked into orbit with a low point of 868 miles and a high point of 11,135 miles, scientists of JPL (Jet Propulsion Laboratory) decided to maneuver their bird about to settle it down *just* where they wanted it. After tweaking the thrusters they kept the low point of orbit at 868 miles, but shifted the high

THOSE INCREDIBLE ROBOTS

point down to 10,655 miles. They shifted the inclination of orbit (to the equator) from 64.28 to 64.36 degrees. When they were through skidding and yawing and rolling the distant machine racing around Mars, Mariner IX was right smack where they wanted it so that at the low point of orbit (periapsis) the spacecraft would be directly in the middle of the "transmission view period" of a huge antenna at Goldstone, California, to get the best possible results for both telemetry data and television pictures.

Immediately that Mariner IX was locked into orbit, American scientists notified the Soviet government of the orbital details for any effect it might have on the Russian spacecraft, Mars 2. On November 27, as it approached the planet, Mars 2 ejected a small capsule. A retrorocket fired and dropped the capsule on a collision path with the Martian atmosphere. Not very much has ever been learned about this capsule except that it carried metal banners and emblems and was the first probe ever to reach the Martian surface. Did it carry radios or instruments? No one really seems to know, and the Russians to this day have remained strangely silent on the subject. Soon after the capsule was ejected, Mars 2 went into its orbit about the planet. Low point was 860 miles and high point soared to 15,600 miles for a looping elliptical orbit. Mars 2 swung around the little world inclined 48.54 degrees to the equator.

On December 2, Mars 3 came upon the scene and ejected a landing capsule that left no question as to its purpose. The retrorockets fired on time and the landing robot slammed into the atmosphere, taking high-g loads and absorbing the heat of entry on a jettisonable heat shield. Once it slowed down, the Lander released a small stabilizing parachute, and followed with a large chute. Just before impact, more rockets fired downward to cushion the touchdown—a system long used by the Russians for returning manned vehicles from orbit, as well as automatic probes coming back from the moon.

The Mars 3 Lander dropped to the surface in the southern hemisphere of Mars between the Electric and Phaetonic regions, pinpointed at 45 degrees south latitude and 159 degrees west longitude. The Lander had aboard scientific instruments, radio and TV camera transmitters. According to the Russians, between December 2 and 5 signals were received from the Lander by the Mars 3 automatic station in orbit. The television transmissions, for which there had been tremendous anticipation, lasted only twenty seconds, and went dead. No one knows what happened. A dust storm was howling on Mars. Did the Lander tip over? Did sand cover the lens? Was there a malfunction of equipment? All these questions and more will have to wait until a manned ship reaches Mars 2—just as Conrad and Bean of Apollo XII finally reached Surveyor I on the surface of the moon.

One of the more enigmatic aspects of these flights is that, while we know the Russian probes carried out an extensive instrument survey of Mars, we know very little about the results of their photographic sessions of the planet. Mariner IX would produce a staggering number of pictures, but the only ones the public ever saw taken by the Russian probes were grainy and crude. Did the Russians get clear photography? No one seems to know. Sometimes, frustrated scientists will tell you, the mystery of working with other men can exceed the mystery of studying other worlds.

Mariner IV took twenty-one pictures of Mars.

Mariners VI and VII took two hundred pictures.

Each photographic session threw the scientific world into a quandary, because each time they studied the pictures returning from the distant world of Mars, they saw cherished theories crumbling to ashes.

But, with the enormous amount of information available from these three photographic probes, we thought we knew what to expect from that point on. Mariner IV had photographed just one percent of the red planet. Mariners VI and VII increased the coverage to twenty percent of Mars.

Mariner IX, over a period of eleven and a half months, with

THOSE INCREDIBLE ROBOTS

698 revolutions of Mars, provided a map of the entire planet. And wiped out all the *new* theories that emerged from the pictures of the previous Mariners!

The miracle of sending thousands of pictures across several tens of millions of miles, from one world to another, down through a thick and turbulent atmosphere, is difficult to appreciate, for the simple reason that it's so difficult to understand. It's not something you explain in a few words. You've got to take it slowly, and, even so, we can only pick out the highlights.

For our purposes we'll select Mariner IV—the *first* deep-space robot to transmit pictures between worlds. And if we keep in mind that Mariner IV took only twenty-one pictures, and later Mariners took five, ten, and a thousand times that many, we can better understand the enormous advances that were made in spacecraft technology in the brief span of time from one mission to the next.

The most surprising first thing to learn about a robot spacecraft like Mariner IV is that it is so light in weight—only 575 pounds. It would take a couple of Mariners to add up to the weight of a small sports car. Within this lightweight wonder there are more than 138,000 components of different sizes, shapes, weights, alloys and purposes. The separate electronic components were made up of 31,696 parts and pieces, and these went into a dazzling array of systems that ranged from a compact electronic brain to a radio transmitter of ridiculous capability. I say ridiculous, because who would ever expect a transmitter with an operating power of only 10.5 watts to be able to send photographs, as well as scientific and engineering data, across a distance of *135,000,000 miles*? It seems so impossible it's laughable, but it's just what Mariner IV did.

Engineers took this intricate and sensitive contraption and subjected it to years of tests: cold soak, heat sink, high-g forces, weightlessness, spinning, vibrating, shaking, bouncing—you name it and they had already thought of it. Then they folded its gleaming solar cell wings, filled its small fuel tanks for thrust control, gave it a final pat on whatever goes for a spacecraft backside, and snuggled it tightly within a clamshell adapter atop a multistage rocket booster. When all was said and done, and the planets balanced neatly along the plane of the ecliptic, their positions measured down to millionths of a second, they launched the rocket. *More* vibration, from the rocket and from the resisting atmosphere. Then grueling g-forces from acceleration. Crash into weightlessness as the first stage drains its tanks and is kicked away like so much gleaming garbage. Bang! The second stage fires and Mariner booms into Earth orbit. Weightless again until everything is checked out, and Mariner slingshots around the Earth and the engine ignites—-and crash! Severe acceleration, a howling ride to more than 25,000 miles an hour, shutdown, and back into zero-g. A couple of jolts here and there as explosive bolts kick the precious payload free of its rocket stage. Weightlessness again.

Now the real mission *begins*. Mariner IV has nine months to go in an environment just about guaranteed to break it down. There's solar radiation that heats exposed metal and other materials to hundreds of degrees. Anything not exposed to that radiation is wickedly cold. Then there's interplanetary dust, micrometeoroids, the debris of space. Solar storms send raging torrents of particles into the space through which Mariner must sail. Cosmic radiation streams through the mechanisms. Through all this, Mariner must do more than survive; it has to sustain itself by recording the details of its environment and also by studying how its own machinery acts. It draws in energy from the sun on its cells, converts this to electricity, translates scientific and engineering data to coded electronic signals. It makes sure its sensors point to certain stars and planets, it adjusts its balance along an invisible frame of reference by thrusting with its tiny jets, and then it must "lock" onto an antenna millions of miles away and continuing to recede with every passing minute, and *then* transmit its information. It also receives new instructions, compares them with what's been packed inside its small computer, and de-

THOSE INCREDIBLE ROBOTS

cides what to do next. And this is long before it ever gets near Mars to start its photographic mission . . .

In the vicinity of Mars, sending information back to Earth is so extraordinarily difficult it seems beyond all possibility. Remember the small size of this gleaming robot and the distances involved. Space is filled with horrendous static and interference in the electronic sense. Radiation storms, magnetic storms, gravity waves, cosmic radiation and other lines of interference must be overcome. When you have to shout (in the electronic sense) across a distance of 135 million miles, you do so carefully and slowly. You send your message with infinite care so it may be picked up and understood at the receiving end.

Mariner IV, remember, is a fragile mechanical-electronic insect skittering through space. Back on Earth three antennae, each eighty-five feet in diameter and located in California, Australia and South Africa, wait to pick up the signal from the spacecraft. Well, "pick up the signal" is clearly the most understated remark in the space age. Talk about sorcery and miracles—here it is all about us.

Mariner sends its signal out with a 10.5-watt transmitter. But that's across 135 million miles of space. Very quickly, after leaving the spacecraft, it hollows out to a ghostly whisper. Moving with the speed of light, at more than 186,000 miles *per second,* it still takes that tiny, hoarse cry twelve minutes to reach Earth after it leaves Mariner.

By the time it gets down to us and gasps through the atmosphere it's so faint it doesn't seem to exist any more.

The signal is reduced to .000000000000000001 of one watt.

That's eighteen zeros—one-tenth of a billionth of a billionth of a watt.

That's just a little less than one quintillionth of a watt.

Well, how do you pick up something that's almost not there? More sorcery. As the signal comes into an antenna receiver, it's got to be captured so it doesn't get lost, and, after we know it can't escape into whatever lies beyond this sort of between-worlds whisper, it must be brought back to something we can handle with some sense.

Signals are piped through liquid helium in the form of a cryogenic (supercold) maser. This boosts them to a signal strength a thousand times greater than when they reached the antenna. But it's painfully and excruciatingly slow. It takes nearly nine hours to transmit *one* picture from Mariner, 135 million miles away, back to its patient, supersensitive antenna.

The limited power of Mariner cannot transmit a picture across this distance. It does what's within its capability, which is to translate the picture. It sends what is really an equivalent of the pictures being snapped by the TV cameras sliding past Mars. How does it send an equivalence-photograph signal?

Back to the numbers game—mathematics. It converts the captured visual scene (the photograph) to a mathematical expression of the photograph. And to do this it uses a digital form that is based on a binary code. What it all means is an intricate back-and-forth translation of a printed picture into mathematical bits and pieces.

Each picture made up by Mariner is itself made up of 200 lines. Think of the picture this way: two hundred lines starting at the top down to the bottom. Okay. We know the picture is made of those 200 lines. Next question: What makes up each line?

Aha! Dots. That's right. Little old familiar dots. Two hundred dots to each line. So we have a picture made of 200 lines, each of which is made of 200 dots. Spelled out another way, each single photograph is made up of 200 dots across and 200 dots down, for a total of 40,000 dots for each picture. Those are our bits and pieces. Scientists dislike the term dots and prefer to call them elements.

Why? Among other things, the dots/elements must be broken down to still smaller units. The scientist starts with 40,000 elements. He breaks each element down into binary bits—six bits per element. Now, each binary bit is a complete unit of informa-

THOSE INCREDIBLE ROBOTS

tion that will be recognized as such by the antenna and the computers on Earth waiting to receive them.

Computers are really superfast morons with a billion toes on which to count. You and I would quickly go mad trying to deal with the binary bits for a single photograph, and it would take us years just to *count* the pieces that go into a single picture. But the computer? Day and night it happily counts on its electronic toes.

One single picture, starting at the bottom, is made up of binary bits, six of which equal one element, of which there are 40,000 elements per picture; there are 200 elements per line and there are 200 lines—and we end up with 240,000 bits of binary code for a single photograph.

Nearly a quarter of a million pieces and chunks. But, up to this point, our information is still in analogue form, and we've got to convert it to digital language. An electron beam within Mariner scans a photograph in analogue. This gives it an electrical analogy of the light intensities detected by the TV camera. The electron beam reacts to the intensity of light from each dot. (Ever enlarge a picture *way* up? It *is* made of dots . . .) The beam then converts the light from each dot into a numerical code that runs from one through six. Things get a bit sticky when you learn that each individual dot of any picture can be any one of sixty-four shades of gray, starting at the bottom with white and going all the way to black. After the dot (taking them one at a time) is converted to its code of six numbers, the numbers were then sent by radio back to Earth. Here they were recorded and run through a computer. The computer simply noted the numerical code of each dot, and printed pictures made up of dots of varying grayness based on that numerical code.

Well, not really. The computer was not printing a picture. All it did was to print dot after dot in line after line and *men* saw a picture. The computer didn't have that distinction.

Each dot is made up of six binary bits. Mariner IV transmitted with a speed of 8.33 bits of data per second—and when you

PLANETFALL

need to handle 240,000 binary bits for each picture, you see why you need nearly nine hours to assemble the numbers into a single photograph.

Each Mariner improved upon its predecessor. Mariner IV weighed 575 pounds. It had one subsystem to send back scientific and engineering information with a speed of 8.33 bits per second. During its mission it sent back to Earth 325.5 million bits of such information, two-thirds of which were scientific data, the remainder engineering information.

Mariners VI and VII increased the spacecraft weight (each) to 910 pounds, and they had three subsystems each for data transmission. Not at 8.33 bits per second—but at 16,200 bits per second. They also sent back two *billion* data bits in scientific information and one billion data bits for engineering.

Then came Mariner IX with a weight of 2,150 pounds when it reached Mars—with half this weight made up of the fuel to slide into Martian orbit, and for attitude control during its long mission. Mariner IX increased the science return to *49 billion bits.* Wait—there's more. Mariner IX was intended to operate for ninety days around Mars in its primary mission.

Three months. How do you evaluate the success of a spacecraft that, instead, operated for eleven and a half months and mapped an entire planet?

It makes up, we might say, for whatever failures went before.

And it gave us a planet of which we were in almost complete ignorance—a fascinating, living, dynamic world.

Time to visit the "new" Mars.

Chapter 13

The Mars We Never Knew

The ocean of air that girds the planet Mars is thin and naked. Compared to the heavy atmospheric seas which swirl about our own globe, Mars has been stripped of much of its gases by a weak gravity and by time. On that bare and arid world, most if not all of its surface condemned to oxygen starvation and sterility, the distant sun blazes with a harsh and pitiless glare.

There are no mountains on Mars . . . even the largest hills of this desert planet do not lift their tops to more than several thousand feet above the surface. Deadly monotony rather than exciting variation dominates the Martian landscape. It is a world bleak and bare, covered by the spilling fall of reddish-brown and yellow sand, sprayed with finely powdered soil . . . when the gales of Mars shriek fiercely, that cuts with needlelike force through the naked atmosphere.

There are devastating sandstorms on Mars. When the storms march, they roar like giant walls of dust moving across the vastness of a planetary desert. Much of the sand bulks up into crescent dunes or leans across the surface in rippling patterns and waves. From six to nineteen miles above this surface, there are to be found thin and wispy clouds, colored a strange blue-white, and most likely composed of frozen particles of ice.

Well, that's what I wrote about Mars in *The Greatest Challenge* ten years ago, basing the descriptions on the latest information we had obtained from telescopes and radar studies of the distant planet, and reflecting the opinions and conclusions of scientists who had spent their lives trying to wrest from the mysterious and fascinating world the secrets it had managed to keep from us so long.

At least I was right to *some* extent. The Martian atmosphere, now that our robots have been to the red planet, has been confirmed—and that is a *very* different word than anticipated—to be thin and naked, stripped of much of its gases, and with little oxygen remaining. Certainly I was right about the howling sandstorms on Mars (duststorms might be a better word for it), for when Mariner IX and Mars 2 and 3 reached the planet, a storm, in its last stages, had hurled with furious energy a mantle of sight-obscuring dust that covered the entire world. And you can't get a storm much bigger than *that*.

I recall, when I wrote that description of Mars, that more eyebrows than I care to remember went up with the description that "Much of the sand bulks up into crescent dunes or leans across the surface in rippling patterns and waves." A small crowd of friendly experts ridiculed that extrapolation. Whatever dust or sand there might be on Mars, they said, could hardly be moved about like that. Not piling into sharply defined dunes. The air was too thin. The winds might raise duststorms, but they'd be more like clouds than anything that could create large dunes.

Well, score one for me. I don't know *why* I was so convinced there *must* be dunes on Mars, but I was. I'd spent time in the Sahara Desert, have walked through White Sands and the Mojave and half a dozen other deserts, and it impressed me that sand can be ground down to a lot less than the grains of sand we find on the beach, and this gives them a powdery nature, and almost *any* wind is going to blow such material about. If the wind is strong and it blows from a consistent direction, as it does at times on Mars, you get dunes. One of the Mariner IX pictures that I found so delightful, then, was taken of the floor of a crater ninety-three miles in diameter in the Hellespontus region. Ninety-three miles is a respectable chunk of real estate, and this crater, from one rim to the other, is a magnificent field of high, sharply defined dunes.

There was one more for our side, where I had written of clouds "most likely composed of frozen particles of ice." I confess that possibility dimmed with the survey of Mars by Mariner IV, faded even more after the missions of VI and VII, but came back with wonderful justification when the instruments of Mariner IX left no question that some high clouds of Mars *were* of frozen ice particles. Then I felt a touch of jubilation when it was shown that "extensive cloud systems" were of water-ice crystals.

Now, what about the statement that there "are no mountains on Mars . . . even the largest hills of this desert planet do not lift their tops to more than several thousand feet above the surface." To add insult to injury I continued: "Deadly monotony rather than exciting variation dominates the Martian landscape."

Wrong is hardly the word for it. The only consolation is that the whole world was wrong right along with me. They were wrong even *after* the pictures of Mariner IV. They were wrong even after the two hundred pictures from Mariners VI and VII. They were wrong even after the first pictures of Mars were taken from Mariner IX before it was drilled neatly into orbit. Because not only did Mars turn out to have mountains, it had the biggest, most incredible mountain we had seen on *three planets.*

Before we went to Mars—through our robots—our Earthbound telescopes had taken pictures of a mysterious feature on the planet's surface that was named Nix Olympica. It was a strange bright spot that sometimes waxed and waned in its brightness. No one knew what it was, except that it gave birth to a bewildering variety of equally bewildering theories. Everyone seemed to join the act. Nix Olympica was a great sea of Mars reflecting light. Nix Olympica was a huge oasis. It was a system of great mirrors the Martians were using to reflect light in their attempts to signal the inhabitants of Earth. It was a great crater. It was a sheet of ice that covered hundreds of miles that accidentally reflected sunlight and seemed to be a mirror surface.

PLANETFALL

Mariner IV didn't shed much light of its own on the subject because it wasn't in a position to photograph Nix Olympica. Mariners VI and VII got some beautiful shots of Nix Olympica—a strange, very bright ring formation. A huge *crater* three hundred miles across. Mariner IX began to collect some more great pictures of the strange ringed formation, and then scientists began to assemble close-up photos of the area, and they were stunned with what they saw.

After detailed photographs of the planetary surface taken by four Mariner spacecraft, and only *after* these pictures and *after* careful study of the pictures, the assembly of the close-up photos showed that *every previous picture had been misinterpreted.*

Nix Olympica was the greatest volcano known on three worlds. It was a volcano with a diameter of 310 miles—wider than the entire state of Missouri. It was a volcano with a peak reaching to 48,000 feet above the surrounding plain!

Do you remember the Mars we once knew?

The world with huge canals?

With darkening waves of vegetation that followed the change of seasons?

With immense artificial waterways?

With a moon, Phobos, that seemed so strongly to violate natural orbital laws that serious thought was given to the fact that Phobos might be an enormous artificial satellite?

That had mysterious bright flashes some scientists suggested, very seriously, might be the explosions of thermonuclear bombs?

A world that was absolutely dead?

A world host to many forms of life?

Ad infinitum.

Well, there's no need for gross speculation any longer. Mars is incredibly fascinating. It is so completely different from anything we knew or expected that it might as well be a planet of another solar system.

First, some specifics. The diameter of Mars is just about half

THE MARS WE NEVER KNEW

that of Earth—4,216 miles. As an interesting sidelight, Mars is about twice the diameter of the moon, and Earth is twice the diameter of Mars. The mean distance between Mars and the sun is 141,650,000 miles, and the red planet swings through an eccentric orbit with a minimum distance from the sun of 128,330,000 miles and a maximum of 154,760,000 miles.

Mars receives just about half the radiated energy from the sun that Earth does. If you stood on the surface of Mars and looked at the sun (through a dark filter) you'd see a star only two-thirds the size of the sun we know on our home world.

Mars's gravity is quite close to that expected—about thirty-eight percent of Earth value.

The surprise is that the Martian day is unusually, even astonishingly, close to the one we know—or 24 hours 37 minutes and 22.58 seconds. Another striking similarity is that where the Earth tilts from the plane of the ecliptic at an angle of 23.5 degrees, Mars is almost precisely the same—24.8 degrees in its tilt.

Mars orbits the sun with a speed of fifteen miles per second (compared to Earth at 18.5 mps). Each Martian year lasts for 687 days—almost twice that of ours. That's 687 Earth days—and 668 Martian days. The northern hemisphere has a spring lasting for 199 days, a summer of 182 days, an autumn of 146 days, and a winter of 160 days. Weather on Mars is extremely responsive to position in that eccentric orbit. Closest to the sun, Mars receives forty-five percent more radiated energy than it does at the greatest distance, producing climatic effects so severe the changes are easily noticed by Earth telescopes.

Now for the world itself—and there is so much that is new (just about *all* of Mars is "new")—we'll move along briskly in spanning the planet. First, as indicated by Mariner IV and confirmed by later spacecraft, Mars is heavily cratered. At first the craters seemed so dominant that scientists believed they were looking at a larger model of our own moon. The craters were of every size and shape, although they had one characteristic never seen on the moon. They were weathered, worn down, beaten up,

185

aged. Thin atmosphere, ice, duststorms of severe intensity—all these factors contributed to the aging characteristics.

It might be better if we took our tour of the planet at slower speed. So, for the moment, we'll stick to what we learned about Mars through the flights of Mariners VI and VII, which took some two hundred pictures, representing about twenty percent of the Martian surface, as well as making scientific observations of conditions on the surface and within the atmosphere. They showed the atmosphere to be approximately ninety percent carbon dioxide. Atmospheric pressure at the surface depended upon elevation. Pressure on the surface of Mars is the same as the pressure on Earth 100,000 to 125,000 feet up. There seemed to be water vapor of puzzling quantities, in the form of ice crystals and "waterized atmospheric fog." There was ionized carbon dioxide, carbon monoxide, atomic hydrogen and traces of molecular oxygen in the atmosphere. White clouds appeared suddenly in the polar regions with the coming of summer.

There was ice, and there was snow on Mars, although Mariners VI and VII left unanswered the question whether the snow was frozen carbon dioxide or was originally water. Craters with snow or ice several feet deep along the crater floors and rims came out clearly in photographs.

These Mariners, in covering only twenty percent of the planet, showed that Mars was made up of three specific types of terrain that dominate the surface. Keep in mind that before Mariner IV we had no idea that Mars would be heavily cratered. The pictures of only one percent of the surface threw the scientific world into a turmoil. Scientists were both shocked and stunned to realize they had a glimpse of a world totally strange to them, despite years of studies from Earth and all those years filled with every possible conclusion drawn from their observations.

Then, after the dust settled down in the meeting halls, Mariners VI and VII blew the whistle again. This time the reaction was of overwhelming silence. After all, how often can you keep on being shocked on the same subject?

THE MARS WE NEVER KNEW

The most prevalent feature (again, as indicated by Mariners VI and VII) was cratered terrain. Many craters were thirty to forty miles in diameter, and one was three hundred miles across (this was Nix Olympica, still believed to be a bright ringed crater). Areas pounded with craters showed up heavily about the south polar cap. The equatorial regions also showed heavy cratering (and, in retrospect, it seems astonishing that these two spacecraft should have missed the most dominant features on the planet, which would be picked up by Mariner IX). Very quickly it became clear that comparing Mars to the moon because of the craters could be a serious error in judgment. Craters on Mars were plotted by size, depth, slope and other characteristics, but they differed greatly from comparative plots of our moon. The Martian craters were being eroded by weather.

Featureless terrain came as one of the bigger surprises, especially when the craters photographed by Mariner IV had brought scientists to expect more of the same, keeping in line with the dominant characteristics of the moon. But of many areas, photographed with a resolution down to a thousand feet in diameter, all that could be seen was a featureless, empty surface. Especially in the bright desert floor of Hellas was this evident. Over a distance of 1,200 miles, the sharp cameras of the two spacecraft could not pick up a single feature of a thousand feet or larger. Obviously, there might be craters of smaller size, but the point was that there *should* be the larger craters—and they just weren't there. The enigma could be solved only by accepting that Hellas was being subjected to much more severe erosion processes than elsewhere.

The terrain distinctions made it necessary to remind those receiving this new information, when referring to certain areas of desert, that *all* Mars is a desert—based on pressure, temperature, humidity and other factors. Or, at least, a desert by Earth standards.

After the cratered and the chaotic terrain, many scientists were convinced that Mars had sprung whatever surprises it had

PLANETFALL

in store, but then came, once again, completely unexpected features. The chaotic terrain, which appeared only in two narrow-angle pictures from Mariner VI. Many times the wide-angle pictures showed gross detail and gave no hint of the extraordinary features hidden by distance. Then the narrow-angle cameras, zooming in for a tight shot, brought forth the incredible and totally surprising terrain. Thus Mariner VI whetted curiosity more than satisfying it. The pictures, stated the scientists controlling the spacecraft mission, showed "irregular, jumbled topography, reminiscent of the slumped aspect of a terrestrial landslide. . . . In a specific Martian area of about 772,000 square miles, as much as 386,000 square miles may consist of chaotic terrain."

Temperature studies taken as the twin spacecraft swept past Mars showed there could be no single general reading for the planet but that temperatures depended upon time of day and season and upon the ability of the surface materials to absorb or reflect solar radiations, the latter in itself affected by clouds. Mariner VII recorded temperatures along the boundary of the south polar cap at 193 degrees below zero F. During daylight, noontime temperatures along the equator climbed to 62 degrees above zero F. Along the edge of the polar cap they climbed from minus 193 at night to minus 45.

One of the key points of this mission was to shed some light on the famous "wave of darkening" on Mars so often seen by astronomers on Earth. Let the official report tell that one: "There is no evidence of seasonal darkening in the television pictures; no distinctive topographic or physiographic character has been revealed by the pictures in the regions where the observed 'wave of darkening' takes place."

To which many scientists said "Phooey." Twenty percent of a planet wasn't enough to satisfy them. Not at all. They'd wait for the next missions.

There were indications—and this is 20-20 *hindsight,* obviously—of just how wrong conclusions can be that are based on

THE MARS WE NEVER KNEW

eighty percent of the planet being unseen. Mariners VI and VII led to the conclusions that atmospheric clouds or fog were essentially restricted to the polar regions. This would be shown to be drastically in error.

Most Martian craters failed to indicate the central peaks so common to the moon. Neither were there the rays or streaks that span out from the lunar craters, but this was hardly surprising in view of the atmosphere, the winds and erosion of Mars. Also, men have walked directly through or along the lunar crater ray streaks and have been unable to determine any change about them. This has turned out to be a feature noticeable by the gross reflection of light, rather than pointing to physical alterations of the terrain.

The twin Mariners showed low and irregular ridges on Mars, rather similar to those seen on the moon. But what about straight or sinuous rilles? None appeared in the two hundred pictures. But they would be there when Mariner IX went to work . . .

And who could anticipate—what had escaped the lenses of the twin Mariners—monstrous volcanos, huge shallow pits, gigantic canyons, rilles, mountains, vast lava flows, calderas, convective cloud formations, water channels and great stores of water-ice hidden beneath the surface!

You have two moons. Both small, both close in their orbits to the parent body. One of these moons doesn't behave according to everything we know about celestial mechanics. If our observations are right, and if our mathematics are faultless, the body we're observing *must* be an artificial satellite. Nothing in nature explains its behavior.

That alone would make the trip to Mars worthwhile. For beyond question the two Martian moons, Phobos and Deimos, are among the most unusual satellites we know. We had always believed Phobos to be about ten miles in diameter, and Deimos half that size. What they turn out to be are giant chunks of rock, rather than our impression of a moon, such as our own natural satellite with a diameter greater than two thousand miles. Before the Mariners arrived at Mars, we had learned enough about reflectivity of spatial bodies, through our lunar studies, to alter the estimated size of Phobos to fourteen and Deimos to eight miles.

Standing on Mars, Phobos appears in the sky with a diameter only one-third Earth's moon (as we see our satellite). Deimos is cut down to even less than that, showing in the sky as Venus does from Earth.

Phobos has a circular orbit only 5,800 miles above Mars, and it really hustles along, whipping through a complete orbit in seven hours and thirty-nine minutes, only one-third the Martian day. Here was Clue Number One in the mystery of Phobos, for the period of revolution of Phobos, shorter than the rotation of the planet on its axis, is stated by astronomers to be "an absolutely unique phenomenon in our solar system."

Deimos is the smallest known moon in our system, and it orbits Mars at a distance of 12,500 miles, with a period of thirty hours.

Thus the two moons behave in peculiar fashions, again drawing special attention. They're described as "dynamical nightmares." They don't behave according to the rules.

Let's say Phobos rises (in the *west,* because of its enormous speed) as a thin crescent. It whips along so fast that before your eyes it changes from that thin crescent to a full moon! Eleven hours after you saw it rise, it would have gone completely around its world and would be soaring again above the horizon. Deimos plays its own tricks. Since its revolution is close to planetary rotation, it seems always to be trying to catch up to the surface below, always slowly losing the race, and therefore hanging in the sky during any one appearance for two and a half Martian days.

American and Russian scientists were joined in their chagrin about not understanding the characteristics of the tiny moons of Mars. Baffled might be a better word for it. Soon after the moons were discovered in 1877, astronomers drew up the orbits, calcu-

THE MARS WE NEVER KNEW

lated for years to come. This works out very well in celestial mechanics, and almanacs are filled with such information.

They weren't correct very long for Phobos, which astounded astronomers by ignoring the rule book and moving in its orbit by 2.5 degrees which, in astronomical terms, is both heretical and impossible. Russian physicist I. S. Shklovsky points out that in "just a few decades, Phobos moved as much as 2.5 degrees away from the point in orbit, where, according to calculations, it should have been. An incomprehensible fact—simply a scandal in celestial mechanics!" There was more. Phobos was dropping toward Mars and increasing its orbital speed. We are, states Shklovsky, "witnessing the slow agony of a celestial body. It means that, in just a mere fifteen million years, Phobos will fall in on Mars."

Astronomers speculated that dust might be causing the orbital decay, but since Deimos was locked neatly into orbit, that idea was dropped. What about a major magnetic field? Too outlandish. Shklovsky stated: "There are no natural ways to explain either the origin of the Martian moons or the oddities in the movement of Phobos."

The theories began to snowball. Phobos was dropping because of outgassing of the thin Martian atmosphere. But to do this it had to be hollow, with an average density of only 1/10,000th that of water. So it couldn't be a solid body. If not solid, what then? "Phobos is hollow inside," insisted Shklovsky. It was impossible, then, for it to be of natural origin. "Consequently," concluded the Russian scientist, "Phobos is of artificial design . . . an artificial satellite of Mars."

More than a few American scientists went along with this possibility.

Until Mariner VII and Photograph Number 91. From a position 86,000 miles above Mars the cameras picked out a speck above the light region known as Aeria, just west of Syrtis Major. The speck was enlarged and run through computers to enhance its detail, and the mystery surrounding Phobos went up like a puff of smoke.

The moon was shaped like a giant potato, a lopsided chunk of rock eleven miles from pole to pole, and fourteen miles in diameter at the equator. It turned out, also, to be the darkest body ever seen in the solar system.

Mariner IX locked into orbit about Mars and started taking its pictures. The incredible space robot sent back a total of 7,329 photographs, the last one transmitted from a distance of 238,416,000 miles.

Several of those pictures were of Phobos and Deimos. The latter also was irregular and lumpy, and both moons were smashed and battered by the impact of meteoroids. Phobos had one enormous crater in its side; speculation that the changing orbit of Phobos was due both to such terrible collisions *and* the outgassing of the Martian atmosphere seemed to be settling onto solid ground.

There remained one great surprise to astronomers. Both Phobos and Deimos were confirmed to be synchronous with Mars—one side of the moon always faced the planet, the other side always faced away.

The "mystery" of the Martian moons was ended. But whatever the Mariners had taken away in putting flight to the idea of artificial satellites, they more than made up, through Mariner IX, in the revelation of a world almost completely strange to men.

Mariner IX, after 698 orbits of Mars over a period of 349 days, ended its astonishing year of operations with the close of an age of speculation and the opening of a door to a new world, now guaranteed the intense interest of the intelligent life forms that inhabit the planet known as Earth. For Mariner IX did more than take pictures and sniff out electronically the characteristics of the red planet. It remained operational long enough to study Mars through more than half a Martian year and was thus able to survey the planet below as the seasons went through their changes. The mission of Mariner IX ended on October 27, 1972. Although it would take many months more to break down the information obtained, some of the major observations and conclusions, stated in their simplest form, were enough to close the door on the yesterday of painful guesswork.

Mars is confirmed as a geologically active planet with volcanic mountains and calderas larger than any ever known on Earth.

There is an equatorial crevass—the Coprates rift valley—that runs for 2,500 miles in length. It is like the Grand Canyon but on an incredible scale. It is seventy-five miles wide and twenty thousand feet deep and covers a distance of one-quarter of the circumference of the planet!

There is no longer any question that free-flowing water was present in the surface of Mars in the past.

The duststorm encountered by Mariner IX and Mars 2 and 3 when they arrived in the vicinity of the planet was hurled up in only three days' time, raged to an altitude of thirty-five miles above the surface, and covered the entire planet.

Cloud cover is extensive on Mars and includes convective (vertically building) as well as layered clouds. The duststorms and cloudiness of the planet are clearly the case of the puzzling variability in the appearance of Mars as it has been observed for many years from Earth.

The duststorm encountered when Mariner IX arrived had been going on for two months before the spacecraft went into orbit, and it lasted for six weeks more. In many ways it proved a scientific boon, when scientists discovered that the effect of large amounts of dust in the atmosphere was to cool the surface and warm the atmosphere—*a measurement of enormous value to scientists who have long been trying to calculate the effect of increasing pollution on Earth's global climate.*

During the initial part of the mission, while the colossal duststorm covered most of the globe, all that was visible through the dust to Mariner IX was the bright, waning ice cap at the south pole and four dark mountain peaks barely discernible through the thick haze. Later photographs showed that each "dark spot" was a giant volcanic mountain with a large caldera-type crater system at its peak. Nix Olympica, the largest, was 310 miles in diameter at the base and loomed nearly eleven miles above the surrounding plain.

Mars was discovered—through alterations to the orbit of Mariner IX—to have severe gravity field variations.

By January 1972, as the duststorm cleared, Mariner IX looked down on "a brand-new target, a clear Mars." The pictures that followed revealed an unexpected world. The south polar cap was shrinking rapidly. A vast system of sinuous channels was found, and the channels were estimated (and later proved) to have been slashed out of the surface by flowing water. Chaotic terrain was photographed in detail. Huge impact craters were found, with their floors covered from rim to rim with windblown dunes.

On January 12, Mariner IX took detailed pictures of a 300-mile section of the enormous Coprates rift valley and discovered branching canyons that had eroded the adjacent plateau lands.

The immediate postmission science summary showed that Mars, after the entire surface had been photographed, is subdivided into at least four major geological provinces. First is the Nix Olympica-Tharsis volcanic province. Second is the Ophir-Eos equatorial plateau region with extensive faults and rifts. Third is the area containing cratered and smooth terrains, perhaps more ancient than the first two types, found in both the northern and southern hemispheres. Large circular basins (Argyre I and Hellas) bear close resemblance to impact basins on the moon. The fourth subdivided area is the south polar cratered terrain blanketed by glacial sediment layers up to 400 feet thick. Similar deposits appear in the north polar region. "If large quantities of water exist on Mars," noted the official report of Mariner IX, "they are undoubtedly locked in the permanent polar caps."

Once the duststorm subsided, photographs showed that much of Mars above 45 degrees north was covered by a north polar hood of variable clouds. Special attention was paid to gaining expanded coverage of such features.

Details of the surface features piled up in dizzying contradictions of everything believed about Mars before Mariner IX. There were the volcanoes and the rilles and the craters and—well, no one expected the planet to *bulge*. No one really expected Mars to have a bulge about 110 degrees west that stands about a mile and a half above the surface; on the other side of the planet the bulge is repeated. And between the bulges there are areas slightly depressed along the equatorial zone of Mars. Scientists had expected some gravitational roughness but nothing so severe that it would quickly alter the expected orbital path of Mariner IX.

Craters, of course, were expected, but not dense clusters of craters, thirty to sixty-two miles in diameter, so thickly concentrated that they resembled the chaotic uplands of the moon.

Slowly and steadily there was revealed to us a fascinating world of huge craters and volcanoes, huge pits and depressions, rilles, lava flows, low mountain ranges split right down their centers, incredible canyons, all manner of cuts and slashes and drifts across and through the surface. And many of these features were, geologically speaking, of recent origin. Mars was *alive*.

The rate of change in surface features brought on by windblown dust caught everyone by surprise. When Mariner IX completed taking its first 1,500 pictures, scientists began comparing the same area photographs that had been covered by Mariners VI and VII. Many of the craters pictured in the early photographs had been *almost completely filled in by dust*. Here there was no lunar surface "frozen in time."

On December 17, 1971, from a height of four thousand miles, the spacecraft snapped a picture of Phoenicis Lacus, an area only three weeks earlier covered with dust. Now the picture revealed a series of mountain-high ridges, making up a terrain scientists described, in amazement, as "wrinkled like an elephant's hide." In an area about seventy-five miles square there were impact craters, indicating that this particular area had been like this for a long time. There was another surprise to come. The upthrusting wrinkled surface loomed to eighteen thousand feet higher than the surrounding surface.

"Extraordinary" was the name for strange pits and hollows that showed up in the south polar regions, for these features were like "deflation hollows" seen on Earth. On Mars, they are likely the results of wind erosion and melting ice and snow.

Another photograph emerged from a bonanza of riches, covering an area 336 by 264 miles, showing a dramatic and startling network of canyons up to fifteen miles across each separate feature. Immediately the unusual area was dubbed the "magnificent chandelier" by scientists, and was identified as dramatic evidence of erosional processes at work on the fractured volcanic tablelands of Noctis Lacus.

Then, of course, there were more and more pictures taken of the Coprates rift valley. Compare the Grand Canyon of the United States—217 miles long, eleven miles wide and one mile deep—to the Coprates rift that is 2,500 miles long, seventy-five miles wide and four miles deep!

The first sign of possible water erosion in the past of Mars showed up in a valley of Tithonius Lacus, about three hundred miles south of the equator. Plateau walls rimming a down-faulted valley fifty miles wide . . . with erosional patterns exactly like those of Earth.

Still the details poured in, and scientists finally had the chance to settle back, study and match the findings of instruments against the incredible photographs.

Under the north polar cap there exists an enormous cache of water that holds the potential of providing to Mars an atmosphere as dense as that on the Earth.

The information began to mount. The photographs that had been studied were run again and again through an enhancement process in which computers brought out details with vivid clarity that seemed to unlock doors only unhinged before this moment. First there was the evidence of glacial fields deep beneath the polar cap, and the realization that if great amounts of energy could be brought to Mars, this could be released into the atmosphere. For the volcanoes of Mars were already pouring a hundred thousand gallons of water into the atmosphere every day!

The enhanced photography showed beyond any question that running water, some of it like floods brought on through melting glaciers, in the tropic zones of Mars, had cut huge channels in the surface.

At least nine major volcanoes were photographed, and the four largest had extensive cloud systems in their vicinity. The clouds were photographed hugging the lower slopes, and *may* be caused by outgassing from the volcanoes. If this is the case, then beneath the surface, in that area, the deep ground could be hot and volatile and on the edge of continued eruption (which every scientist would dearly love to see).

Of particular interest is that the volcanoes and intense crustal faulting are fairly close together on the Martian surface, suggesting that Mars remains dynamic and in the throes of planetary change. Some scientists feel they have learned enough through Mariner IX to suggest that there may be vertical convection currents well within the interior of Mars and that the crust undergoes the counterpart to the continental drift of Earth.

Ozone was found in the high atmosphere, peaking during the winter and almost disappearing during summer months. Even though the heaviest concentration of atmospheric ozone above Mars is but two percent of that found above Earth, it is considered a major scientific find.

Some erosion was known to exist because of dust/sand and high winds, but Mariner IX established conclusively that wind erosion is a dominant feature in the past and present history and activity of the planet. In middle latitudes, during the huge global storm that started before the arrival of Mariner IX, speeds of sixty-five to ninety miles an hour seemed to be common, and winds up to 110 miles an hour were estimated by visible effects on the surface. The Martian dust also left huge streaks, considered a result of sand-size particles laid down by the winds that point to a pattern of global air circulation.

If the atmospheric circulation truly is global it means that Mars is undergoing convection, where air rises into the upper atmosphere because of a heated surface below. As the air ascends it cools, spilling outward in descending currents over cooler regions, beginning a widespread pattern. Most scientists are absolutely convinced there is convection *and* a global weather pattern, and they emphasize that the Mariner IX photos have shown both stratus (layered) clouds and convective, cumulus types.

If the clouds about the lower slopes of Martian volcanoes are not caused by outgassing (and they could be both, of course), then most certainly they are the result of convection over heated ground and uplift along the slopes.

Equally exciting is proof that the clouds are formed of water-ice crystals.

Piece by piece, more characteristics of Mars emerged. The water-cut channels are real. They happened at different times in the past history of Mars. They happened so long ago that several of the deeper channels show heavy weather erosion.

If the Earth were as far from the sun as Mars is, *almost all the water on this world would be frozen solid.* Reversing things, it becomes clear through many signs that Mars has enough water frozen beneath both polar caps so that, if at some time in the past, the eccentric orbit brought Mars closer to the sun, vast quantities of water would be released as melting ice and as rain. There are four huge volcanoes—Nix Olympica and three others on Tharsis Ridge—that alone could have enough water and vapor from their craters to form a substantial Martian atmosphere, which would bring about drastic changes in temperature and pressure.

How old is Nix Olympica? Geologically, it's a young volcano. There are no craters along its slopes.

Throughout the entire planet, Mariner IX photographed unmistakable lava channels that resemble sinuous rilles seen on the moon (especially Hadley Rille explored during Apollo XV). The lava channels, scientists confirmed, are entirely different from those specified as having been cut by flowing water.

The height of the Martian volcanoes is considered "amazing." Middle Spot Volcano on Tharsis Ridge (near Nix Olympica) is eight and a half miles above the Ridge, and the Ridge itself is about nine miles above adjacent basins. The volcanoes also provided an unexpected boon when they were first photographed. Their young geologic age, and the clarity of the Mariner pictures, shows that Nix Olympica has long lava channels and ridges sluicing its slopes, an indication of basaltic (watery) lava flow. But South Spot Volcano on Tharsis Ridge shows a slope pattern with short and stubby channels and grooves. Scientists point out that this indicates a differing chemical composition of the lava flows, which for South Spot are more viscous and higher in silica and alumina than the basalts. The closest comparison to terrestrial materials would be the adesitic rock types found commonly in the Andes Mountain chain of South America.

Thus Mariner IX during its lifetime has been extracting from the world below a wholly new portrait of a planet we never knew —not shrouded any longer in mysteries of ignorance, or hidden behind the veils of far distance. Now we have been given a world of enormous energies, alive and different and, above all, challenging.

The world toward which huge new robots will soon be speeding, and after which men will follow, as surely as they followed their robots from the Earth to the moon.

CHAPTER 14

Mercury to Mars— the New Robots

There's not very much to say about Mercury, and for a good reason. We don't know much about it, and it's frustrating. Since Mercury is close to the center of the solar system, the first planet outbound from the sun, a few words and figures about our star might be in order. It provides a nice scaling effect that reminds us of just how small and seemingly insignificant is the world on which lives the race of man.

One way to look at the sun is the numbers game that tells us our star is 864,000 miles in diameter. A paltry number unto itself to describe this incredible sphere of blazing nuclear fire that every second converts 564 million tons of hydrogen into 560 million tons of helium, in the process neatly annihilating four million tons of hydrogen. It gets rather heavy when we realize the sun has been carrying out its process of annihilation for perhaps five billion years or more—five billion years of annihilating hydrogen at a rate of four million tons every second.

Try and add up *those* figures, and when you're through, think about stars such as Aldebaran, which every second annihilates 640 million tons of matter! Then think of the red giant known as Epsilon Aurigae; it is so unspeakably huge that it would require *27 billion of our suns* to equal this one star in size.

And then place in your mind, taking it nice and slow, that the sun makes up 99.86 percent of all the matter in our solar system. Doesn't leave very much, when you think about it, and way down there near the bottom of the scale is our Earth.

MERCURY TO MARS—THE NEW ROBOTS

It's a dense planet. The densest of all the planets, so far as we can tell. And then think of Jupiter, which has a mass 1,300 times greater than Earth—a mass more than twice as great as the other eight planets combined.

Okay, back to the sun. Or rather, let's go far beyond the sun to a distance of five billion light years. I can't conceive of what it really means and neither can you, but at least it's a numbered yardstick. Within this distance from our sun, according to the best estimates of astronomers (which are changing all the time and seem to have less accuracy than trying to predict next year's weather), there are 100 billion galaxies.

Each galaxy is an overwhelming family of stars—containing 100 billion or more of these glowing suns.

Now, let's work our way through to just one galaxy. There's no use trying to picture 100 billion galaxies, each with 100 billion (or more) stars, because we can't do it. There's no use even trying to comprehend the 100 billion stars that make up the Milky Way Galaxy. But we *can* attempt a visualization of the Milky Way by imagining we can see it from a distance so great that the entire galaxy falls within our vision. And what we see, if we're looking down upon the entire galaxy, is a glowing, blazing pinwheel, with a heavy concentration of stars near the galactic center. The outer streams of stars of this fiery pinwheel are in the shape of glowing spirals.

We've got reference points, and since our vehicle is imagination, we can move just about as quickly—and anywhere—as we wish. So we drop into that seething mass of stars clustering about the bulbous shape of the galactic center, and start rushing outward from the center toward one of the spiral arms. As we hurtle along, the stars begin to thin greatly in number until we're in one of the lesser neighborhoods. Still the number decreases, and soon we're reaching the outstretched spiral arm along the outskirts of the galaxy. The galactic center, about which everything orbits, is now 180 million-billion miles away.

We're home. We've just arrived at our own familiar, "friendly" star—the sun. It moves at a snail's pace through the bottomless pit of space, a measly 482,000 miles an hour. At this speed it requires a mere 220 million years to complete one orbit about the galactic center.

But it's a neat star, at that. It's the most perfect sphere in the solar system, although it's far from being a really perfect sphere, because it's slightly flattened at its poles. Poles? The *sun?* Absolutely, although it's rather a complex and involved movement. One complete rotation at a point of the solar equator takes 25.4 Earth days. But near the solar poles, thirty-four days are needed for a full rotation, so that part of the sun always seems to be hurrying to catch up with the main body near the solar girth.

Not *that* far from the sun is a small and dense body racing in its orbit at 110,000 miles an hour. It's a sort of rogue little world, defying the general law of symmetry that "calls for" the planets to orbit generally in a circle. Mercury, aptly named for its orbit speed, at times cuts along in its elliptical path around the sun to a distance from the star of only 28 million miles—and, at the far end of its orbital dash, manages to get to 43 million miles from the sun.

Every eighty-eight days (all "days" are given in Earth time), Mercury completes a full revolution about the sun. Until recently —and in astronomical terms, the year 1965 is *very* recent—Mercury was "known" to present the same side to the sun, in what we call a synchronous orbit (like our moon or the twin moons of Mars). It seems of particular interest that until the last few years we were convinced that *both* Venus and Mercury matched their period of rotation and revolution.

According to all the rule books by which we understand celestial mechanics and the motions of bodies in space, Mercury *should* have been synchronous and should have always kept the same face toward the sun. Averaging only 36 million miles from the huge star, Mercury is subjected to massive gravitational forces that eons past "should have" canceled out all independent motions on the part of the tiny planet. And so it should have,

MERCURY TO MARS—THE NEW ROBOTS

and, if it did, then one side of Mercury would have been seared and broiled by the blinding solar radiation, while the other side, without atmosphere, would have been one of the coldest spots in all the solar system, being turned away forever from solar heat.

It wasn't that long ago that the supposed dark side of Mercury was discovered *not* to be as cold as predicted. In fact, astronomers were positively annoyed that it seemed to be quite warm. They first learned this disquieting information in 1965, when scientists released some research work details with the thousand-foot-wide Arecibo radiotelescope in Puerto Rico. The instrument had been used to fire short and high-powered bursts of radio waves at Mercury, fifty-seven million miles away. Five minutes later the bursts of energy hit Mercury and bounced, and in another five minutes Arecibo caught the pulsing return signal and locked the signal onto tape. Somewhat more leisurely, scientists then began to dissect the signals, checking the frequency of the returning beams to measure the changes in the signals. What they found was disturbing, because it contradicted all previous beliefs.

Mercury, they discovered, was just about to rewrite all the textbooks. It did not always present the same face to the sun. It completed a full rotation every fifty-five days (give or take five days either way for errors) while it revolved about the sun once every eighty-eight days. So one side was not always hot and the other a celestial deep-freeze.

But certain celestial laws really do stand up to time, *even if we don't understand the signs they give us.* Mercury should not have had independent rotational motion. It did. *How?* Time for the theories. Thomas Gold, director of Cornell University's Center for Radiophysics and Space Research, offered a possibility. In fact he was convinced he had the answer, which said, rather grandly, that Mercury was not a true planet. Everything pointed to the fact that Mercury was once a moon of Venus. It had its own independent rotation, and slowly it drifted away from the parent planet. About 400 million years ago, Gold theorized, it fell under the domination of the sun and whipped into orbit, an orbit that differs greatly from the other inner planets by being sharply elliptical. And Mercury is spinning, all right, but once it was spinning much faster, and the sun is still damping out that rotational movement.

It could be. Meanwhile, because it is so close to the sun and moves so fast, Mercury has managed to evade any real scrutiny on our part. We know that Mercury is just about as dense as the Earth. In fact, if the Earth, with its diameter of 7,900 miles, were to be made entirely of iron, it would have a diameter of 7,040 miles. And Mercury, with its diameter of 3,100 miles? If it were entirely of iron it would have a diameter of 2,410 miles. (By comparison, Saturn has a diameter of 75,100 miles. If *it* were iron, its diameter would be 32,120 miles.)

Some other comparisons show that Earth has a mean density 5.52 times that of water, Mercury is almost with us with a mean density 5.45 times that of water, and Mars is figured at 4.92 times that of water. That's why Mercury, with a diameter about 1,100 miles less than that of Mars, has just about the same surface gravity.

Certain figures and conditions can be extrapolated. Standing on Mercury, an observer would see the sun more than twice the size of the fiery orb we see from Earth. And Mercury receives just about five times the radiant energy on its surface that we know here, so that the temperatures can only be described as searing. We've learned a great deal about the planets through our close observations of the moon and of Mars. Mercury, under the blazing assault of the sun, and because of its own light gravity, has no chance of retaining an atmosphere. If, as we believe, the solar system at times much earlier in its formation was as filled with debris as the moon and Mars indicate, then Mercury must be a cratered world.

It is also a world of hellish conditions. The radiated energy falling on Mercury is five times greater than we know here on Earth, but that's only the average. When Mercury dips in its orbit to come closest to the sun, that energy level *doubles* to ten times

MERCURY TO MARS—THE NEW ROBOTS

that experienced on Earth. The best estimates for surface temperatures range from 600 to 900 degrees F., which has certainly melted or vaporized most mineral deposits. There's every chance that when Mercury takes this awesome battering it has a temporary atmosphere: metals flaming and melting to release heated vapors that slowly outgas into space. There might even be a cloud of vaporous material whipping along in orbit with the little planet.

We may soon increase our knowledge of Mercury by a hundredfold—just as Mariner IX in a few months changed drastically all concepts held about Mars. There's always a danger in writing about what's going to happen; at least there's a danger today, because so many stupendous events are taking place all about us.

For instance, engineers and scientists are preparing a number of great rocket boosters and their payloads for new missions away from Earth. One of those vehicles is scheduled for launch on November 3, 1973—an Atlas-Centaur. When the bird stands on its pad, within the great nose cone will be a 1,000-pound spacecraft named Mariner X.

And if the launch goes as planned, then Mariner X will race by Venus at a distance of 3,300 miles, and bend in its flight to aim toward Mercury, to slip by the innermost world of the solar system at a distance of only 620 miles.

Aboard Mariner X will be the largest telescope ever built so far for planetary exploration. Think of the miraculous accomplishments of Mariner IX, and imagine now what we can learn with the vast improvements in the spacecraft systems for Mariner X to Venus and Mercury.

"If all goes well"—which is a tribal chant here at the Cape—then Mariner X will transmit back to Earth no less than 4,500 separate vidicon frames, each with a resolution that will average out to the best pictures received from Mars.

In one stroke—again—the mysterious veils of distance and inaccessibility will be lifted.

There are two television cameras aboard Mariner X, and they

PLANETFALL

are set to take a total of 8,000 pictures of both Venus and Mercury.

The mission will begin at Pad 36B at Cape Canaveral, the rocket booming upward into Earth orbit. Swiftly, details of the flight are checked out, and then the Centaur upper stage is reignited to reach escape velocity from Earth. (There is also the chance that, depending upon the actual time of launch, the flight will take place without a coasting orbit—the Atlas and Centaur stages will continue burning steadily to send Mariner X on its way.)

Aboard Mariner X will be six scientific experiments, other than the two television cameras and transmitting equipment, to return both planetary and interplanetary data. But the big emphasis is on Mercury, and the management team—the Jet Propulsion Laboratory of the California Institute of Technology—is highly experienced in its work, having directed the JPL Mariner 1969 and 1971 programs to Mars.

Every small change in a booster and spacecraft system has possible large effects on a mission, and Mariner X is to be sent into space aboard a new-model Centaur known as D-1. It will look like earlier Centaurs, but the difference is beneath the skin, since the D-1 model has a guidance system highly improved over the Centaurs fired earlier (which did a fabulous job on the Mars missions).

Specific areas of interest have long been mapped out by the JPL scientific team, and, hopefully, Mariner X will provide answers to the long-sought questions. What is the precise shape of Mercury, and how does that shape relate to the planet's motion?

What is the nature of the surface morphology and what forces have acted to shape it?

What is the surface temperature of Mercury and how does it vary from local night to day?

Is there regional color and variation in brightness of the surface, such as we have found on Mars and the moon? Is there any

MERCURY TO MARS—THE NEW ROBOTS

atmosphere? If so, of what density and what composition? Is it being replenished?

How strong is the magnetic field, and what is the nature of its interaction with the solar wind, this close to the sun?

Then there's Venus . . . and the chance to get the first pictures of that planet, as well as to get more details to add to what's been learned from the Russian probes that descended into the atmosphere. Is there an observable vertical or horizontal or circulation pattern in the visible clouds of Venus? What is the form and motion of mysterious ultraviolet markings that have been noted in that atmosphere? Are there any holes or breaks in the thick cloud cover that may permit looking deep into that atmosphere?

To be successful the first double-planet robot exploration requires what is known as the "gravity whip." In this gravity-assist flight plan, Mariner X comes close enough to Venus to be accelerated by the attraction of the planet, and then have its trajectory modified—or bent—by that gravitational field. Venus will bend the trajectory of Mariner X and accelerate it so greatly that the spacecraft will pick up enormous velocity and "whip" toward a new trajectory that will bring it close to Mercury.

Launch is set for November 3, 1973. Three months later, Mariner X will be making its flyby of Venus.

On March 29, 1974, if the flight plan works out as intended, Mariner X will begin its initial encounter with Mercury.

Included in the experiments are an infrared radiometer and an extreme ultraviolet spectrometer to carry out thermal mapping of both planets, and to conduct atmospheric studies of the two worlds.

Mercury is the major goal of the mission, and the encounter is broken down into three broad periods. During the encounter the spacecraft will be transmitting information in both real time and from tapes of instrument findings.

The first encounter phase—*approach*—starts seven days out from Mercury and lasts until the spacecraft is only one day away.

PLANETFALL

At this time the planet will fill the entire narrow field of view of the long focal-length camera system. Mariner X will be able to resolve objects down to twelve miles in size at this point.

The *near* encounter begins at sixteen hours before the closest approach, and ends four hours before closest approach.

Encounter includes the four hours before and after closest approach.

During the approach to Mercury the pictures taken by the cameras will be taped for later broadcast. At encounter, minus four hours from closest approach, real-time television will start, and Mercury's portrait will be flashed back to Earth just as it is being seen by Mariner X. The greatest number of pictures will be sent in this real-time (as it happens) photography, since it takes forty-two seconds just to record on tape one entire picture frame. Taped pictures and information will be transmitted before closest approach and afterward, during times when the spacecraft would be looking at the dark side of Mercury. The highest resolution will show details of Mercury's surface as small as 328 feet in diameter.

It should be quite a ride—and only the first with Mercury. Because there's a bonus in the flight plan. After the first pass of the planet, Mariner X will swing out in space, away from the sun, well beyond the orbital path of Venus. It will then bend again in its trajectory—and 176 days after the first encounter with Mercury, Mariner X will again make a close pass to the planet's surface!

Every 176 days from then on—barring any unknown factors that may be introduced by the sun—Mariner X will race by the innermost planet of our system. Hopefully, the second, third and fourth passes will produce more information and pictures. Mariner X carries enough on-board fuel to use its attitude thrusters to aim cameras at Mercury, and antenna to Earth, for six encounters, but most scientists believe the instruments will not survive that long so close to the sun.

* * *

MERCURY TO MARS—THE NEW ROBOTS

And Mariner X will be only the first of the new robots to slice inward of Earth's orbit toward Venus and Mercury. Plans are under way to release—in a 1975 mission—a small balloon filled with hydrogen or helium, into the Venusian atmosphere. There it would float for days, weeks and perhaps months, carrying beneath it an instrument package that would take long-time and detailed explorations and studies of the atmosphere.

Waiting in the wings to make its debut is a twin-Pioneer mission to Venus in 1978—a demanding multiprobe and orbital robot expedition that will require two separate launches.

Scientific teams for the 1978 Venus mission are being made up from both the United States and several European nations. As the planning continues, NASA has decided upon the Atlas-Centaur D-1 booster for each launch. Estimates at this time indicate the payload will be on the order of just under one ton.

Two launch opportunities are available in 1978—one during May-June, and the second in August.

The first shot would send three instrumented probes into the Venusian atmosphere, and a larger vehicle to be lowered by parachute. Then the spacecraft—called a bus—would also enter the atmosphere for scientific measurements.

The second launch, following the first by about ten days, would send the orbiting probe toward Venus with a very high velocity, in order to achieve a high-inclination or even polar orbit, which would drop at low point to 126 miles and a high point above the planet of 38,000 miles. Once in orbit, the probe would begin gross-feature radar mapping of the surface.

Beyond the twin-Pioneer mission, scientists are planning a major flight either in 1981 or in the 1983-1984 period. This would be known as the VOIR (Venus Orbiting Imaging Radar) probe. The VOIR vehicle would carry extremely advanced radar mapping systems that would be able to make out details as small as 40 feet in diameter.

A special report on studies of Venus notes: "The Mariner experience with Mars shows that complete areal coverage is as essential as adequate resolution. Although Mariners VI and VII photographed about 10 percent of the planet at a resolution of several kilometers (about 2 miles), the sample was extremely misleading. Conclusions based on the scant Mariner VI and VII coverage have been in many cases reversed by Mariner IX pictures."

There will be another major change in the exploration of the planets, especially Venus and Mars. The governments of the United States and the Soviet Union have agreed upon a major exchange of information on what they have learned about these worlds, so that each will gain the benefit of what has been accomplished by both countries. The initial agreement, which came in early 1973, calls for the Russians to provide the United States with full data from its Mars 2 and 3 spacecraft, so that the United States will have every possible scrap of information to assist in choosing the landing sites for two Viking robots intended to touch down on Mars in 1976. The U.S., in return, will provide the Russians with the detailed photographs and maps that resulted from Mariner IX's mission to the red planet.

Much more information will be exchanged, such as:

Projections based on data obtained from Earth radar on the location and movement of Mars in the first half of 1974. This information was sent to the Russians by NASA prior to a deadline of April 15, 1973.

The United States will also send to the Russians radar measurements of Mars, all information obtained on Venus from the mission of Mariner X, and whatever significant new information is obtained on either planet.

On their part, the Russians will provide this country with full information on the atmospheric measurements they made of Mars and the surface of Mars. Especially important is the Russian breakdown of the atmosphere as it pertains to a landing ship, which is needed for our Viking flights. The Russians—in terms

MERCURY TO MARS—THE NEW ROBOTS

of Venus—are providing us with complete information on Venus as determined by Venus 8 and their reports on radar studies of Venus.

The United States plans Mariner X as the first of their new series of Venus-bound robots, to be followed by the 1975 mission that will eject a balloon into the Venusian atmosphere. The late-in-the-decade missions include building each flight around a standard-shaped spacecraft, known as a "bus," to reduce the complexity and cost of these flights.

Two Pioneer missions to Venus are now scheduled (after the 1973 and 1975 launches) to begin in January of 1977. The first spacecraft will eject four highly instrumented probes into Venus's atmosphere. About sixteen months later, another ship would be sent to Venus, and placed into orbit about the planet, as Mars was orbited by Mariner IX, to carry out extensive long-period studies.

Still further into the future is another Venus probe to go into orbit, but this time with a highly advanced radar system to map the entire planet. To be launched about 1985, the probe would go into polar orbit and then use its radar systems to obtain a map of Venus during a period of 243 Earth days. It would be able to register objects as small as a thousand feet in diameter.

But there's also Mars . . .

Mars 3 was the first spacecraft to descend to the surface of the red planet, and its future distinction will certainly be the first of a long line of robot visitors to thump down on the sands of Mars. Seeing how Mars 3 operated provides us with an excellent look into the future for those ships to follow.

The Russian Mars 3 Lander was made up of a hermetically sealed instrument compartment that contained scientific instruments, a rotating television periscope, and radio and telemetering instruments. Attached to the Lander's exterior were special sensors designed to be jettisoned after the robot touched down on the planet. These would make direct-contact studies of the surface materials. The findings would then be radioed back to the Mars 3 Orbiter, which would in turn retransmit back to Earth. All mission requirements were to be attended to through an elaborate time-sequence generator.

The Lander also was fitted out with a conical heat shield with special fins to radiate away heat, retrorockets, parachutes and special explosive systems to separate the various parts.

The Lander separated from the mother ship while eight hundred miles from Mars. It coasted for fifteen minutes, then the Lander engine ignited. The robot turned about to enter the atmosphere with the heat shield-braking cone in forward position. Four and a half hours after leaving the mother ship (that remained in orbit) the Lander began to feel the effects of Mars's atmosphere. It was still plunging at supersonic speed when a drag chute whipped away. This provided both deceleration and stability, and then the main chute boomed out, remaining in furled position to reduce opening shock. When the speed eased off to just about Mach 1, the speed of sound, the main chute opened fully. A radar altimeter measured the changing distance to the ground through the descent.

Seventy feet above Mars the final braking rockets flashed. The instant they burned out a single rocket ignited. This blew away the parachute to prevent any interference with the Lander or its antenna after touchdown. The Lander dropped the final few feet to the surface and came to a stop. Immediately petal-like stabilizers (used successfully with lunar probes) opened to steady the Lander.

Mars 3 was on the surface in the southern hemisphere in the area between Electris and Phaetonis, with coordinates of approximately 45 south latitude and 158 west longitude, in what is considered to be a plains area with relatively few high surface features. Smooth or not, Mars 3 transmitted only briefly. No one knows if it struck a rock, fell into a crater, was tumbled over by the winds and dust—or what. It transmitted its video picture for twenty seconds and went dead.

MERCURY TO MARS—THE NEW ROBOTS

But that it landed successfully was not in doubt, and the type of information the Russians had gained was of enormous value to the United States for its upcoming Viking program. Both American and Russian scientists believe the Lander struck ground with a speed of 68 feet per second, surviving a short-pulse impact load of 500g—only about half the load it was designed to take and still function.

The descent, or landing, rockets fired when Mars 3 was about ninety feet above the surface, dropping at 275 feet per second. The fiery blast of 10,000 pounds thrust effectively slowed the spacecraft.

Total time through the atmosphere had taken three minutes.

The Russians made the first descents to the surface of Mars, and their Mars 2 and 3 spacecraft returned an undetermined number of photographs of still questionable quality. The United States had lost Mariner VIII, but came up roses with Mariner IX and its extraordinary photographic-scientific performance in orbit about the red planet.

No one on this side of the ocean, however, believed for a moment that the Soviets were going to yield an inch in their determined effort to become kingpins in Martian exploration, and in late July of 1973, they ended all speculation on that subject.

On July 21, Mars 4 boomed away from Earth, whipped into low orbit, and, before one revolution had been completed, gashed the vacuum night with renewed flame and was on its way to the red planet. But the window—the period of time in which energy requirements for a Mars journey were at their best—was still open. Four days later, on July 25, Mars 5 followed. The Soviets didn't reveal much, but it was significant. They had used an SL-12 booster, and, apparently free of the myriad problems that had plagued so many of their shots to Mars, were enjoying great success with the two probes. Little information as to the payload was released, and the Russians murmured that the weight of Mars

PLANETFALL

4 and 5 was approximately that of the Mars 2 and 3 missions fired in 1971.

But the Russians had been playing it close to the vest, and on August 5 they caught everyone napping with a third salvo in their barrage to the fourth planet—Mars 6 slammed into orbit and reignited its upper stage engines for a beautiful slingshot maneuver toward Mars. Eyebrows were raising over the outstanding booster success of the Russians; Mars 2 through 6 made it five in a row.

On August 9 the Russians did it again as Mars 7 cracked the centrifugal whip and raced toward the fourth planet from the sun. Now the pattern was clear. Mars 4 and 5 were expected to reach their destination in late January 1974. Mars 6 and 7 would show up on the scene about five or six weeks later. By then the Russians would have accurate orbital plots of the first two probes—and would be ready to place 6 and 7 right where they wanted them.

Moscow revealed in another tidbit of information that Mars 6 and 7 "differed somewhat in construction" from the early two probes. They also stated that 4 and 5 were analogous, and 6 and 7 also were analogous to one another. In other words, two sets of twin spacecraft had been launched—two to orbit Mars and two to land. Apparently the new probes were considerably heavier than Mars 2 and 3, when the orbiters and Landers flew together to the red planet.

What about the payloads? The Russians are concealing their cards. Estimates from this side of the ocean are that at least one of the two Landers is to carry a Mars Rover—a planetary edition of the Lunokhod multiwheeled vehicle that trundled about the moon. The Russians, a few years back, when discussing their rovers for Mars, had stated the command system would be computer-based and part of the installation sent to that planet. This, of course, is the major difference between the lunar and Martian rovers; the close-to-home version on the moon is commanded by a ground station in Russia.

MERCURY TO MARS—THE NEW ROBOTS

It should be interesting, the spring of 1974. Mariner IX, Mars 2 and 3, and then Mars 4, 5, 6 and 7 all in orbit at the same time —as well as the vehicles on the surface.

That is, in the old tribal chant of the space age, *if all goes well* . . .

We'll have to wait and see—you, me and the Russians.

And the group most intensely interested, and waiting impatiently, is known as *Project Viking*. They're keeping one eye on the Russians, and with the other they're studying two specific areas on Mars.

The first is known as Chryse, and it is 19.5 degrees north of the Martian equator, lying in the center of a smooth plains area, charted as being away from the Mars surface where high winds are known to blow. On July 4, 1976, on the two-hundredth birthday of the United States, a wide-legged robot will drop gently to the surface of Chryse. The first Viking will have arrived.

The second area is Cydonia, where a second Viking is scheduled to descend on August 23, 1976. Cydonia, in the Martian lowlands, at 44.3 degrees north and 10 degrees west, is 18,000 feet below the main elevation of Mars. It is a smooth, mottled plain with rolling features. Surface details include fragments of volcanic rock and wind-blown debris, but it is considered smooth enough for the second Viking robot, which will be a thousand miles from the Chryse landing site.

Cydonia is where our scientists believe there is enough water and pressure to support some form of Martian life.

The pressure, because of the low elevation, is heavy enough for liquid water to exist in this particular area.

If life indeed is found . . . Well, your own imagination can tell you the explosive effect this will have throughout our civilization.

Two Viking spacecraft—each consisting of an Orbiter and the Viking Lander—are scheduled to be launched during a 44-day window that opens on August 11, 1975. They will ascend from the Titan IIIC launch pads on the northern edge of Cape Ken-

PLANETFALL

nedy Air Force Station (Cape Canaveral), north of the pads from which Mariners have left for Venus and Mars, and generally southeast of the pads from where men went to the moon.

The Vikings will require almost eleven months to reach Mars, during which time they'll cover a total distance of about 460,000,000 miles. Each booster will be a Titan-Centaur, a combination of the Air Force's most powerful rocket and the NASA deep-space Centaur. Each booster is 160 feet tall with four stages and will weigh about one and a half million pounds at launch.

Spacecraft weights will be determined by many factors still to be decided in detail. Generally, however, the Vikings will be huge and massive compared to the Mariners we've flown to date. Each ship will weigh about eight thousand pounds, including the Orbiter robot at 5,250 pounds (of which 3,250 pounds are fuel), and the Viking Lander, which weighs about 2,350 pounds.

Each three-legged Lander (looking much like a squat version of the Surveyors that landed on the moon) will contain several television cameras and batteries of scientific instruments.

Launch will be "standard" for a Mars voyage. Each ship will be punched into a parking orbit 115 miles above Earth before being hurled toward Mars at more than 25,000 miles an hour. During the outward flight the Lander will remain completely dormant, its systems checked at intervals by the Orbiter. NASA plans to insert each complete Viking into an orbit about Mars from 620 to 20,460 miles above the surface (taking 24.6 hours per complete orbit). From this "parking position" the long-range cameras will study the Chryse and Cydonia landing sites to be certain the areas present no previously detected dangers to the success of the mission. The "study period" in parking orbit can last as long as fifty days before a landing must be made, and, if necessary, another site can be selected if it proves best to make such a last-minute change.

When the descent is ready to take place, the Lander is separated from the Orbiter. A powerful engine fires to decelerate the

MERCURY TO MARS—THE NEW ROBOTS

Lander and start it down toward the surface. It slams into the atmosphere, attitude thruster jets maintaining exact positioning, and the blow of atmospheric entry is absorbed by an aeroshell descent capsule. Based on what we've learned about the Martian atmosphere, the first indications of deceleration will start about 800,000 feet high, but this will be only the merest tendrils of upper gases. The actual landing program begins when the Lander is only 20,000 feet above the ground, at which point a large parachute is slammed open to decelerate and stabilize the landing system.

A radar altimeter coupled to the onboard computer keeps up a running check of dropping altitude. Just about 6,000 feet high an explosive charge blows away the parachute. Viking falls for another 800 feet, and, just one mile high, terminal rocket engines ignite. They're scheduled to burn steadily until Viking is ten feet above the dusty ground.

The engines shut down. In the low gravity, Viking falls like a heavy leaf. Now the mission *really* begins . . .

Chapter 15

The Outworld

Jupiter is the fifth planet outward from the——
Jupiter is what?
The fifth planet from the sun.
Oh? Who says so?
Why, every astronomy book written! They all say so. There's no argument about it. There's Mercury, Venus, Earth, Mars and then Jupiter. The fifth planet.
But what if Jupiter isn't a planet?
That's ridiculous!
Maybe so. It doesn't change what I just said, though.
But how can Jupiter *not* be a planet! If it isn't it would have to be. . . . Well, if it isn't a planet, what is it, then?
No one is really sure. It has the characteristics of a planet.
See——
It also has certain characteristics of a star.
What?
That's right. Jupiter radiates more energy than it receives from the sun.
But that's impossible!
Nothing is impossible. That's the first rule of our exploration in space.
But Jupiter obviously isn't a star. It doesn't burn like the sun. I mean, it doesn't give off energy the way the sun does——
Because you can't see the energy release in the form of visible light doesn't mean it's not there . . .

*   *   *

Well, we could keep going on for a long time in that vein. But it's true enough. The planet Jupiter doesn't behave like a planet. Except in some ways. It doesn't burn in the manner that a star blazes, although it comes uncomfortably close to doing so.

Thus the question: *What* is Jupiter? The answer must be couched in careful steps, and we would be rather accurate if we describe Jupiter as a planet that is an "almost star."

"If it were only a little more massive," explains Dr. Tobias Owen, of the Illinois Institute of Technology Research, "gravitational contraction would release so much energy that it would turn into a nuclear furnace, like the sun or any other star, and become incandescent."

And that, dear reader, is an almost-star.

What other information do we have about Jupiter? We'll place on a shelf, neatly out of the way to avoid confusion, the designation of Jupiter as a thermal enigma (almost-star), and keep our studies of the huge world in the framework of a planet. Immediately we are overwhelmed with size. Jupiter is the largest of all the planets, with a diameter of about 88,900 miles (the latest approximate figure) along the equator. Its density is only a quarter that of Earth, yet Jupiter has a volume 1,300 times greater than our home world.

Here's where planetary figures can be deceiving. Jupiter has a volume 1,300 times greater than Earth, and its mass is more than twice that of all the other eight planets of the system put together.

But the mass is only 318 times greater than the mass of Earth. So we have a world with a diameter of 88,900 miles compared to Earth at just under 8,000 miles, while the surface gravity of Jupiter is only 2.65 times as great as that on Earth. The more we see of the huge planets the more we come to understand the awesome density of man's home world.

But—the surface gravity of Jupiter is a misnomer. It's strictly a theoretical figure, because we're not certain if there is a surface (as we know the surface of the Earth as distinct from its atmosphere) on Jupiter. It may be swirling gases, a form of interplanetary slush—anything but a surface with which we're familiar. More of that later. Now for some more figures to set Jupiter in clearer perspective.

To escape from the Earth's gravitation field, in order to reach other planets, a spacecraft needs a velocity of just about 25,000 miles an hour (depending upon the height above the Earth when that figure is reached), or seven miles a second.

Another way of comparing the figures is to state that it requires a speed of 37,000 feet per second to reach escape velocity from Earth. To escape Jupiter's massive gravitational chains, a spacecraft would need a speed of just about 200,000 feet per second.

Seen from Earth's surface, Jupiter is the second brightest planet and the fourth brightest object in the sky. Its orbit lies an average of 480 million miles from the sun, with a closest approach to 459 million miles and a maximum separation of 506,700,000 miles. That's far enough, at the medium distance, to reduce the solar energy falling on Jupiter to less than four percent to what we receive here on Earth. And in this ponderous, slow, cold movement, Jupiter requires nearly twelve Earth years in its slightly elliptical path to complete one revolution about the distant sun.

Despite its enormous size, Jupiter is a whirling dervish of a world, the fastest rotating planet in all the solar system, spinning rapidly through a full rotation once every nine hours and fifty-six minutes. Let's stop here for a moment. If you take the time to study different astronomical or space texts, you'll find few that agree on this rotational period. Some sources state nine hours and fifty minutes. Others break it down to hours, minutes and seconds, although this is beggaring the intent of accuracy in figures. Many will simply state ten hours and let it go at that, and, until we have better means of judging, ten hours really will do quite well.

Jupiter's surface is banded and/or striped with an unusual clarity, and so it is possible to follow certain identifiable features we can see swinging along with the planet. But to judge the rotational speed on these features, which likely can be cloud or other gaseous formations perhaps hundreds of miles above Jupiter, is to give the observable speed of those cloud formations and *not* that of the planet along its surface. Which is a nasty feat to accomplish, because we aren't sure just *where* that surface may lie, or even if there *is* a distinct surface. So until they get some better yardsticks, scientists generally accept the figure of ten hours, approximately, for the theoretical surface of Jupiter, for the period of rotation. This means that the "surface" of Jupiter is moving with a speed of 22,000 miles an hour during rotation, along the equator, compared to Earth's equatorial rotation speed of slightly over 1,000 miles an hour.

Unfortunately, we can't just leave one area of inquiry and step nimbly to another that may be well isolated from the point of confusion. The problem here is that they're all, one way or the other, interrelated, and what appears as confusion or question in one area loses none of these puzzling issues by trying to change the subject. It doesn't change. You may step from one square to another, but if you've got a long smear of chewing gum on the bottom of your shoe, it always ends up going along with you.

Astronomers who look at Jupiter can see the banded clouds, the Great Red Spot and other features, and can watch Jupiter turning. Well, when you watch the Earth turn, and you see continents and shorelines and islands, you're watching the Earth itself rotating. When we watch Jupiter, no such thing happens. We're watching what's very, very high in the Jovian atmosphere, and unless you know what you're doing, confusion is immediately your partner. Because different portions of the Jovian "visible surface" move with different speeds. When we look at Jupiter we see the tops of an incredible atmosphere, likely many hundreds of miles deep, perhaps even thousands of miles in depth, and of a pressure level that makes our atmosphere look like a whiff of near vacuum. Venus, of course, does the same, but if our speculation is correct, then even Venus, with 1,300 or 1,500 pounds per square foot air pressure, is far from the awesome pressures of Jupiter.

The more we see of Jupiter the more amazing this incredible world becomes to us. Photographs of the planet seem to be distorted. We look at the moon or Venus or Mars and we see a sphere. We look at Jupiter and we see a squashed ball, and it's not distortion at all. The tremendous rotational speed and the fluid characteristics of the planet (not the atmosphere, the *planet*) produce an enormous bulging at the equator. While Jupiter has a diameter of 88,900 miles, the polar diameter is only 77,000 miles—and that's an almost unbelievable difference of 11,800 miles!

The cloud surface of Jupiter visible to us covers an area of approximately 24 billion square miles.

We noted before that the surface gravity of Jupiter (and keep in mind that this is a theoretical surface) was 2.65 times that of Earth. An interesting sidelight is that the Jovian gravity at the top of its thick ocean of clouds is 2.36 times that of Earth.

Let's get back for a moment to that low density. Some scientists have pointed out the density is so low because we measure the diameter of the planet in such a way that we include the deep and thick atmosphere, and therefore we're getting a density reading that's all wrong. The thing to do, they explain, is measure the diameter of the planet *without* that atmosphere, and we'd come up with a density just about the same as that of Earth. It's a neat trick, all right, with only one problem. No one *knows* the thickness of the atmosphere!

The planet we've seen in photographs is a turbulent, uproarious, swirling mass of clouds, effectively disguising what lies beneath. The surface conditions we can imagine, based on every last scrap of information we've gathered to date, likely include a poisonous (to us) atmosphere denser than metal and with winds of hundreds of miles an hour.

If ever you've had the chance to spend hours studying Jupiter under excellent seeing conditions with a powerful telescope, then

THE OUTWORLD

you know the overpowering fascination of watching this greatest of all planets. It's a dazzling sight, the planet streaked white, pink, yellow and green, all divided by darker brown and red zones. There are irregularly shaped areas of bright green and blue as well as dazzling white patches. The cloud belts run parallel to the equator. Often they show wispy patterns and rough edges. At times it is possible to see pastel hues of blue, brown and pink.

No one knows their origin. The conditions change almost constantly, and this is perhaps a testimony to cataclysmic scenes unlike anything ever known on Earth.

Most scientists believe that Jupiter is made up of a mixture of elements similar to those in the sun or the primordial gas cloud from which the sun and the planets were formed. This indicates there are very large proportions (at least seventy-five percent) of the light gases hydrogen and helium. We have also identified hydrogen, deuterium (the heavy isotope of hydrogen), methane (carbon and hydrogen), and ammonia (nitrogen and hydrogen) within Jupiter's clouds.

The cloud colors and rotation of Jupiter give to the planet a striped or banded appearance, parallel with the equator, with large, dusky, gray regions at both poles. Between the two polar regions are five bright, salmon-colored stripes, known as zones, and four darker, slate-gray stripes, known as belts; the South Equatorial Belt is one example. And everything keeps changing. The planet as a whole changes its hues periodically, possibly as a result of the eleven-year activity cycles of the sun.

The Great Red Spot in the southern hemisphere, a huge oval of such immense size that several Earths could easily disappear within it, is frequently bright red in color. Since 1665 it has disappeared completely several times, and it seems to brighten and darken at intervals of thirty years.

There is wide scientific agreement that the cold cloud tops in the zones are probably largely ammonia vapor and crystals, and the gray polar regions may be condensed methane. The brighter cloud zones have a complete range of colors from yellow and delicate gold to red and bronze. Clouds in the belts range from gray to blue-gray. And there is much more to be seen than zones and belts, for much of the Jovian cloud surface seethes with activity in the form of streaks, wisps, arches, loops, patches, lumps and spots, most of which are thousands of miles in size!

The more you study Jupiter the more incredible, diverse and fascinating the great planet becomes. We have paid so much attention to the closer worlds of Venus and Mars that the true treasure-house of wonders of this solar system has not been given our attention. But today that is becoming a thing of the past. First we climbed the gravity well to the moon and then to Venus and after that to Mars, and now Jupiter itself is within reach of the powerful new machines standing so tall at Cape Canaveral.

Even the circulation of the cloud features has been broken down into elaborately tracked and defined forces on Jupiter. The Great Equatorial Current (the Equatorial Zone) is twenty degrees wide, and thunders about the planet with a speed of 225 miles an hour, faster than the cloud regions to each side of it, acting like a magnificent jet stream, like those of Earth but on a scale that dwarfs this planet. Another feature well known to astronomers is the South Tropical Circulating Current that sweeps completely around the Great Red Spot.

The atmospheric pressure at the *top* of the clouds is estimated to be ten times as great as sea level on Earth—or nearly 150 pounds per square inch. Another mind-boggling thought is that the transparent atmosphere *above* the clouds is estimated at thirty-five miles in thickness.

Among all the planets in this system only the Earth and Jupiter are known to have magnetic fields, and that of Jupiter, derived from study of radio emissions from the planet, is on the order of twenty times the strength of the magnetic field of Earth. A result of this enormously powerful field is that the radiation belts of Jupiter are beyond anything even conceivable a few short years ago. Measurements of the radiation belts of Earth, studied against the magnetic field of Jupiter, lead scientists to conclude

THE OUTWORLD

that the Jovian radiation belts *are a million times more powerful* than those of Earth.

Another fact—in a growing list of staggering facts—is that on Earth we receive more radio noise from Jupiter than from any other extraterrestrial source except the sun. Jupiter acts as a stupendous transmitter of radio signals and noise in three forms. There is thermal noise caused by the temperature-induced motions of molecules in the atmosphere. Decametric signals come from the near-light velocity gyrations of electrons around the lines of force of the Jovian magnetic field. And then there is the grand master of them all—decametric signals that are believed to originate from colossal discharges of electricity (like lightning flashes a thousand or more times greater than the lightning we know on Earth) in Jupiter's ionosphere. There is growing belief that these electrical blasts are so mighty that they affect the closer moons of Jupiter, and that there may even be blazing discharges between the high atmosphere of the planet and the moon Io.

How powerful? A single discharge of built-up electrical potential of the decametric radio bursts is equal to the energy release of several hydrogen bombs!

The temperatures of Jupiter are another startling enigma, for despite the distance from the sun, there are temperature variations and levels that stunned scientists when they first learned of them. The average temperature of the cloud tops appears to be about minus 229 degrees F. But the clouds and the diffuse outer atmosphere of Jupiter present many situations, and the latest radiotelescope studies indicate that much of the outer atmosphere is close to room temperature, with the top layer at 68 degrees F.

The *big* puzzle is that Jupiter radiates away about three times as much energy as it receives from the sun . . .

Jupiter is often described as a miniature solar system with twelve moons. Four of these Jovian satellites are so huge they are larger than our moon, and at least two are larger than the planet Mercury. Largest is Europa, orbiting about 416,000 miles

PLANETFALL

away from the Jovian mother world. It is more than 3,000 miles in diameter and a frozen nightmare of gases and ice, possibly with some atmosphere. Ganymede is as large as Mercury and with a mass twice that of Earth's moon. The innermost satellite is 160 miles in diameter, and orbits with a speed of more than 60,000 miles an hour, only 112,600 miles from the great planet. And to add to the ever-growing list of mysteries, the four outermost moons orbit Jupiter in a direction opposite that of the other known moons.

On March 2, 1972, an Atlas-Centaur with a third-stage rocket lifted from Cape Canaveral's Launch Complex 36. When the rocket burned out, Pioneer X was moving away from the Earth with a speed of 32,000 miles an hour, the fastest man-made object ever flown, and heading for a flyby of Jupiter—and then, set on a course that will take it completely away from the solar system and toward other stars. The TRW-built Pioneer weighs 570 pounds, is nine feet long, carries an antenna nine feet in diameter, and, across the deployed antenna, measures twenty feet. It is the first spacecraft to be powered by nuclear energy, getting its power from four radioisotope thermoelectric generators (RTG's) that power all instruments and control systems.

The awesome journey to Pioneer X places demands on space technology never before made. It became the first spacecraft to pass through the Asteroid Belt that lies between Earth and Mars, it will be the first to investigate Jupiter and its moons, and will become the first artificial object ever to leave our solar system. During the flyby of Jupiter, a distance eight times farther than Mars is from Earth must be traversed for signals. It is so great that ninety minutes are required for round-trip communications. If the flight program continues as intended (and you'll know by the time you read these words), then Pioneer X will be in the vicinity of Jupiter for four days and behind the planet for less than two hours in carrying out its scientific mission.

Pioneer XI was launched in April of 1973 and is now follow-

THE OUTWORLD

ing its predecessor spacecraft toward Jupiter. The difference between the two spacecraft amounts to having options for Pioneer XI, which will depend upon results achieved with the first spacecraft, which is scheduled to make its closest pass to Jupiter on December 3, 1973. Scientists controlling Pioneer XI can set it up to follow the path of Pioneer X, or send it along a different course over the planet's surface, making measurements and taking pictures that will complement those of Pioneer X. Among the options is sending the spacecraft on a gravity-whip swing from Jupiter toward Saturn, to pass by that ringed world in 1980.

By early December of 1972 the flight of Pioneer X was progressing so well that the spacecraft's course was sharpened in an attempt to have it pass behind the orange satellite of Jupiter known as Io. This would permit a measurement of Io's atmosphere—if there is an atmosphere. On September 19, 1972, scientists commanded Pioneer X to fire its thrusters to increase its speed by .745 feet per second. That was enough to alter the flight path so that the spacecraft would arrive at Jupiter 17.2 minutes earlier than scheduled—at just the right time for Io to pass between Pioneer and the Earth. The moon would then be about 330,400 miles from Pioneer.

An idea of the extraordinary control possible with the TRW robot is seen in the course-correction maneuvers. These required twenty-five pulses with the thrusters—fourteen of one-half second duration and eleven of one-eighth second firings. Pioneer X, when the maneuver was completed, was scheduled to make its closest approach to Jupiter at 9:23.5 P.M. EST on December 3, 1973.

Chapter 16

The Rocky Road to Space

Between Mars and Jupiter lies what was long considered the most fearsome obstacle to the exploration of the outworlds—a vast zone of cosmic debris known as the Asteroid Belt. Before the flight of Pioneer X, the asteroid belt was a vast and terrible unknown which many scientists believed would guarantee the destruction of any spacecraft passing through its obstacles. The belt is composed of asteroids, also known as minor planets, or planetoids, believed to be rocky, metallic bodies orbiting the sun in a loose swarm. More than one hundred thousand such objects are believed to inhabit the zone—and this is speaking only of relatively large objects capable of being tracked from Earth. The smaller particles must number in the many, many millions.

The asteroids follow orbits roughly circular and in the same path as the planets moving about the sun. There seem to be about as many theories for their origin as there are asteroids. Most scientists agree there *should* be a planet orbiting the sun between the orbits of Mars and Jupiter; instead, we have the vast swarm of rocky debris as small as dust particles and as large as Ceres, with a diameter of 480 miles. Astronomers have identified and calculated the orbits of 1,776 asteroids, estimating that there are at least fifty thousand of them, from one mile in diameter on up to the size of Ceres. Some of the largest include Pallas, with a diameter of 281 miles; Vesta, 244; and Hygeia, 222. The Asteroid Belt itself is a doughnut-shaped region extending from 186 million to 338 million miles from the sun.

THE ROCKY ROAD TO SPACE

But even more important than the width of the Belt is its *depth,* since it extends about 25 million miles above and below the plane of Earth's orbit. This means it's a barrier to Jupiter and beyond. It's too deep, in reference to the ecliptic, for spacecraft to fly under or over the Belt on their way to the outer space of the solar system. If we had limitless energy (like our projected nuclear-powered *L. Gordon Cooper*), then a spaceship could avoid the rocky road. But we don't have that energy yet, and, until we do, there's no other way except pounding through.

Thus the flight of Pioneer X has already been an enormous success in just getting through the wide and deep Asteroid Belt. Based on the experience of this one spacecraft, it appears that high-velocity, smaller asteroid particles present only a minor hazard to spacecraft. Fingers are crossed that Pioneer XI will have the same good fortune in getting through the solar system's obstacle course.

Scientists estimate that there's enough material in the entire Asteroid Belt to produce a planet with about 1/1,000th the volume of Earth. In fact, the leading suggestion of science is that the asteroids originated from a small planet that was broken up by the massive gravitational force of Jupiter, and the debris spilled out to form the Belt. The other leading contender as a top theory is that the material in the Belt condensed individually from the same primordial gas cloud that formed the sun and the planets. If we can accomplish a flight to the Belt and recover some of this debris, there's every chance we'll have a good look at the records of the solar system during its formation.

In the center of the Belt, asteroids and the smaller particles orbit the sun with a speed of about 38,000 miles an hour. Collision of such objects with an outbound spacecraft, traveling in the same direction, would be at about 30,000 miles an hour. Experience with Pioneer X shows that the greatest danger is not from large particles but from the dust-mote-sized pieces of debris.

Not all the larger asteroids remain properly in their orbit about the sun, for there are renegades among the Belt. These wander in highly eccentric orbits through the solar system; one is Icarus, which swings to within 9 million miles of the sun. Hermes dips to within 220,000 miles of the Earth—much closer than our moon!

The past history of the asteroids is of intense interest to us here on Earth. Mars, for example, lies much closer to the asteroids than do Earth and moon. In the past, scientists believe, Mars underwent a savage bombardment by these great masses from space, at a rate twenty-five times greater than the bombardment that smashed the moon into the cratered hulk we see today. Martian history must, of course, be filled with moments of awesome catastrophes as small mountains howled through the thin atmosphere to punch out enormous holes in the crust—many of them hundreds of miles in diameter.

The Great Meteor Crater of Arizona was pounded into the Earth by an asteroid much smaller than Hermes, which comes within 220,000 miles of this planet. The Arizona crater is four thousand feet in diameter, and it was created by asteroid debris that remained after impact with our atmosphere: deceleration, heating, explosive break-up, all taking place before it struck the surface. Hermes is a mile in diameter, but it has a mass of about *three million tons,* and if it ever struck the Earth, gravity accelerating its own speed, it would tear into the atmosphere with a speed of about twenty miles a second. Any city in the world struck so massive a blow would instantly cease to exist. The effect would in all likelihood exceed that of a hydrogen bomb capable of producing a hundred million tons of explosive force.

Though we cannot yet know what Pioneer X and XI will discover about Jupiter, at least we can identify those broad areas to come under study. Each Pioneer will take about twenty types of measurements of Jupiter's atmosphere, radiation belts, heat balance, magnetic field, internal structure, moons and other phenomena.

And there's another point to make. We already know much about the Jovian world, but we can't explain how, why, or what behind certain things we see, such as the Great Red Spot. Whatever we do know is vastly eclipsed by what we do *not* know. What's hidden beneath those heavy clouds? How intense are the radiation belts? What's the source of the thermal energy pouring out from Jupiter? And so on.

But of all the questions that are being asked, none is so intriguing, so filled with promise, as the question of life on Jupiter. When first we went to the moon the scientific world was split wide open with conjecture on life forms that might exist on that airless world, either native to the moon or a result of "spores" drifting down from space, or contained within meteoric material. The moon, it turned out, was as barren as a slab of steel cleansed by solar fire, and quickly the quarantine of astronauts returning from the moon was abandoned as unnecessary. Mars holds—what? A good, fair, excellent, remote chance of some forms of life? It has had flowing water and rain in its past. The temperature in deep spots still gets high enough, as does the pressure, for water to exist in the liquid state. It has some atmospheric protection, it has outgassing of water vapor and other chemicals from its volcanoes, it has some protection through its ozone layer against raw ultraviolet radiation. Venus is more an enigma, yet within its thick and turbulent atmosphere it may be possible for life forms to have spawned, and not necessarily on the surface.

When we study other planets we do our best to discover unusual situations here on the Earth, and several years ago there came about a change in understanding certain life processes on our world that could have a tremendous impact in searching elsewhere for life. Scientists have long known that our atmosphere is an incredible reservoir for microorganisms. The ocean of air fairly teems with disease bacteria, spores and pollen, but a life form has a dead end if it's biologically inactive and unable to feed or reproduce until it reaches solid ground or a liquid medium. In this case the atmosphere is simply a means of transportation.

PLANETFALL

Not all scientists agreed that organisms floating in the atmosphere *had to be* biologically inactive. Dr. Bruce C. Parker, a biology professor at Virginia Polytechnic Institute, completed research in 1970 that brought him to state flatly that there are colonies of tiny plants and animals (which are known as ecosystems) that live through their entire existence within our atmosphere, and specifically within certain types of clouds. In the process of their existence they produce a variety of chemicals that finally reach the surface through rainfall and even windstorms.

Well, what's possible within terrestrial clouds can certainly be possible within the clouds of other planets.

Most especially Jupiter. Estimates of the depth of the Jovian atmosphere *beneath* the cloud layer show just how very far apart conclusions are about that planet, for they range from sixty to 3,600 miles. And no one can argue that the compositions and the interactions of the gases making up that atmosphere are completely unknown to us. We can assume, of course, that if the atmosphere is deep, then it must be dense. A serious estimate is that if the atmosphere has a depth of 2,600 miles, then the pressure at the Jovian "surface," or what may pass for a surface, would be two hundred thousand times greater than that of Earth, because of the total weight of gas in the crushing Jovian gravity.

That would be nearly three million pounds per square inch. That isn't atmosphere, even in the remotest sense as we know it to be, for at the bottom of the Pacific Ocean the greatest pressure is sixteen thousand pounds per square inch.

Whatever the contention about the depth and pressure along the bottom of the Jovian atmosphere, scientists are in general agreement that liquid droplets—water droplets—exist in that atmosphere. And since the planet is believed to have a mixture of elements similar to that found in the sun, it is also almost certain to have abundant oxygen.

It would be the only planet other than Earth with abundant oxygen in its atmosphere.

THE ROCKY ROAD TO SPACE

Just as important, most of this oxygen has probably combined with the abundant hydrogen to form water.

Reasoning further, it becomes clear that if large regions of the Jovian atmosphere come close to room temperature, then there should be present immense quantities of both liquid water and water-ice.

Jupiter's atmosphere contains ammonia, methane and hydrogen—*it appears to have all the essential ingredients of life.*

The above constituents, along with water, make up the critical elements of the primordial "soup" that is believed to have produced the first life on Earth through the process of chemical evolution.

Based on this evidence, and the known radiation and thermal properties of Jupiter, it is the one planet other than our home world that seems extremely likely to have produced forms of life.

And if Pioneer provides more conclusive evidence that this is so, then it is to Jupiter, more than any other planet in the skies, to which we will turn our attention.

I suppose it's time to fish or cut bait.

We'll find life on Jupiter.

CHAPTER 17

Steamboat Time

By the time you read these pages—if all continues to go well with Pioneers X and XI—the outer limits of the solar system brought to the direct study of man's robots will be the planet Jupiter.

And if all goes well, and then some, that limit will be pushed back so far that the distances will be doubled. Pioneer XI, riding the gravity whip of acceleration caused by a neat pass alongside Jupiter, will be sailing toward—and sometime in 1980 will reach—what is certainly the most beautiful of all the worlds in our solar system—Saturn.

Surprisingly little attention has been given to Saturn, and much less to the worlds that spin their orbits beyond. With good reason, of course. They are so remote, and the robot expeditions to them so expensive, that they simply were pushed to the side by the intense efforts to plumb the secrets of such worlds as Venus, Mars, the moon, and, now, Jupiter and Mercury.

Saturn comes into the picture through the back door, so to speak, a free dividend of Pioneer XI and gravity assist, with Jupiter cracking the whip.

We've learned a hard lesson in planetary expeditions to date. Just about all the planets are mysteries. We *thought* we had them figured out pretty well, but the days of being convinced that the atmosphere of Venus was mainly nitrogen, or that Mars had an atmospheric pressure equal to that of nine miles above the Earth . . . well, all that has gone by the boards with the "jagged" mountains of the moon.

STEAMBOAT TIME

With Saturn, we don't have to be concerned with running into the unknown, because before we ever lift a mile above the Earth, racing to the outworlds, we *know* that Saturn is an enigma unto itself. A world not only with ten moons but with the only moon in all the solar system that we *know* has an atmosphere. Titan, the largest of Saturn's moons, is a satellite that has infected scientists with unquenchable curiosity, for it appears it may have atmospheric and surface conditions much like that on the Earth shortly after it solidified into a planet.

Saturn, sixth world out from the sun, twice as far as Jupiter, is also a world with a mass so light it seems to defy all natural laws. It has something else going for it in the beauty department—the magnificent rings that orbit the planet.

All of which appears to make a convincing argument for study by the instruments of gleaming robots punched away from Earth on pillars of rocket fire . . .

Saturn: dazzling, frozen, ringed, accompanied by its retinue of moons, orbits slowly about a sun so distant that it receives but one percent of the energy that bathes the Earth. Every thirty years Saturn completes one stately revolution about the sun in a journey that brings it as close as 863,700,000 miles, and extends the distance at the far point to 939,570,000 miles. Nearly a billion miles distant from its star.

One of the favored impressions of Saturn is that it is so light it has a density only seven-tenths that of water. If you had an ocean big enough in which to immerse the planet, it would pop to the surface and float like a cork. It has a diameter of just over 75,000 miles, but the surface gravity (if we can define the surface, which causes another problem) is barely more than the one-g of Earth's surface. Scientists have created their own geologic picture of Saturn: an enormous planet of which the hard, dense core comprises only ten percent of the mass, with all the rest made up of gases solidified under great pressure.

The brilliant rings of Saturn have long elicited more response to their beauty than to their composition, but it is the latter that has caused much shouting and table-pounding among scientists. I don't believe there's a single issue of planetary make-up that has been more of a Ping-Pong ball in the matter of changing theories than the rings of Saturn.

The rings extend from six to perhaps fifty thousand miles beyond the planet's surface. Their diameter is rather closely measured to 172,000 miles, but their *thickness* is a point of growing contention among astronomers. Some insist the rings are only several inches in thickness, and they point to the fact that stars can be seen through them. The other side has adherents claiming they're several thousand feet thick. No one knows for certain—and those who insist that what they believe about the rings is fact are making strange noises in a field where almost everything we anticipated about the worlds closer to Earth has proved to be either completely wrong or seriously distorted from reality.

But if the specific figures in terms of thickness are in argument, there's no question that the Saturnian rings form the thinnest and flattest surface in nature ever known to man. A model of the rings of Saturn, for example, would be a sheet of metal in the form of a perfect circle—about one hundred feet in diameter *but no thicker than this page you're reading.*

For many years scientists stated that the rings were composed of rocky debris, that is, pieces and chunks of rock and dust. Then they decided that their studies didn't support this belief, and lo and behold, the rocky debris was banished, and in its place we had a theoretical construction of "snow and icy grit"—a vast accumulation of ice chunks with an outer layer of silvery-white ice crystals. Well, that theory lasted barely more than a few years, and the favorite belief at the moment has us back again with the rocky debris. In this corner the answer is that the rings are there, they're beautiful, and we will *not* know their composition until we get a spacecraft out there for a close look. Until about five years ago all the textbooks stated that there were three separate rings circling Saturn. In 1969 we improved the seeing and photographic capability of our big telescopes, and it was finally ascertained that there really are four separate rings. The two innermost rings revolve more rapidly than the two outer formations. Separated by a distance of four thousand miles, the two outer rings, as seen from Earth, are so close together that they seem to form a single great circular sheet.

The telescope paints a dazzling picture of Saturn, and shows bands of white, orange, green, red and yellow. These slowly change their color gradations and even their shape. At rare moments when the "seeing" is excellent, it's possible to distinguish severe and huge disturbances on the "visible surface" of Saturn. Saturn is, then, something on the order of Jupiter—covered with a massive deck of clouds that seems to be largely methane and ammonia, roiling about, but with drastically less energy than the forces playing across the surface of Jupiter.

The surface temperature is considerably colder than conditions along the edge of the Jovian cloud world; the estimates for Saturn are that the temperature never gets above 300 degrees below zero F.

The biggest news about Saturn, however, is not the planet but the largest of its ten moons—Titan, with a diameter of 3,350 miles, which makes it half-again as large as our moon and just slightly larger than Mercury.

Other moons of Jupiter and Saturn are covered with water-ice. Europa and Ganymede, circling Jupiter, are covered with vast expanses of water-ice that may be similar to frost on Earth. Io and Callisto, also orbiting Jupiter, are believed to bear frost particles, but the evidence is a bit slimmer. (Io will be "visited" by Pioneer XI in 1980.)

Titan, if instrument studies bear up, is a moon that has a fairly heavy atmosphere of methane and ammonia. This was accepted "fact" until late in 1972, when new tools for studying the planets began to change the picture. Using telescopes with infrared sensors, scientists began to learn some puzzling things about Titan. First, it was hotter than it had any right to be. How hot is that?

No one knows yet; the distances are too great for specifics. But it is definitely hotter than the other moons, than the surface of Saturn, and that's enough right there for some deep concentration and more studies. When these studies were made, the conclusions were drawn that Titan may be hot enough to support primitive forms of life. Indeed, studies showed that Titan may well have a surface fairly rich in the kind of organic molecules that existed on the surface of a primitive Earth, and that *did* lead to the origin of life on our home world.

But *why* is Titan warmer than it should be? The best estimates are that there's a greenhouse effect on Titan that absorbs much more of the sun's warmth than is reflected back into space. Well, if there's a greenhouse effect, then there's got to be something else in the atmosphere besides methane, and whatever that substance is, there's got to be more of it than was believed. So scientists began to work out a laboratory creation of what they believed is taking place on Titan.

Sunlight hits Titan's atmosphere and strikes the surface. Infrared radiations are trapped by the atmosphere and cause the surface temperature to rise. This is the greenhouse effect.

Okay so far. But there's a hook in all this. The greenhouse effect can't be caused by carbon dioxide and water vapor (the gases that cause the greenhouse effect on Earth and Venus), because on Titan the temperature never gets high enough to permit the existence of carbon dioxide and water vapor except in a frozen state.

The only gas that meets all the conditions is molecular hydrogen. It can remain a gas at Titan's temperatures, it is an excellent absorber of infrared radiation, and it is also the most abundant molecule in the universe. It's been known for a long time that molecular hydrogen was present in Titan's atmosphere, but the amounts were always considered fractional. Now, scientists believe, Titan may be covered with a dense cloud layer—and we've never seen the surface.

This leads to even more fascinating considerations. The atmospheric soup of Titan must come from *somewhere,* and this leads to unusual—and turbulent—processes existing on and within the giant moon. Those processes are strictly a matter of conjecture and they should be regarded in that vein. *If* the theories are correct, however, then beneath the surface of Titan there may be regions of high temperature, composed of liquid methane, ammonia and water, but not in the condition to which we're accustomed. On Titan these materials would reach the surface in the same way that molten lava bursts forth from volcanoes here on Earth. The liquids reach the surface and evaporate, mixing with the atmosphere of Titan as gases. Another change begins, as ultraviolet radiation from the sun breaks them down into molecular hydrogen and organic compounds.

And again we're led down another path of strong conclusions. Organic compounds of this nature are strongly suspected to be the primordial soup from which life originated on Earth. They also produce a distinct reddish color. And where do we see this strong red?

In the clouds of Titan . . . and the Great Red Spot of Jupiter.

Hydrogen, methane and ammonia are colorless. But the organic molecules created in the active processes of the Titan atmosphere are reddish-brown.

And again down another pathway . . . the gravitational pull of Titan isn't great enough to retain atmospheric gases. But they're there, all right, so this means that they must be supplied to the surface. How? The best bet is for some planetary mechanism like volcanoes, such as exist on Mars. And the Titanian atmosphere is much thicker than the Martian.

Then there's more cause for excitement. Experiments carried out at the University of Maryland indicate strongly that the materials in the atmosphere of Jupiter, *and* Titan, are only a few steps short of living organisms. Scientists produced laboratory equivalents of the Jupiter and Titan atmospheres and then subjected them to electrical discharges, which we know exist on Jupiter and believe also take place on Titan. Among the products resulting from the lab tests just described were small amounts of compounds called aminonitriles. These combine with water to form amino acids, which are an essential part of all proteins. Aminoitriles can also go through a complex chemical process to produce nucleic acids, and these contain the genetic instructions in every living cell.

Another link: The mixture produced in these University of Maryland laboratory test reactions came out red in color.

Among other mysteries, Saturn's moons have long baffled astronomers. Two moons appear to be startingly clear and smooth spheres of pure ice at the surface. And Japetus, 2,210,000 miles from Saturn and half the size of our moon, has one side that is five times brighter than the other!

Saturn, as we have seen, may have its first robot visitor in 1980 with a flyby of Pioneer XI. But there's more to visiting Jupiter and Saturn than with the two probes at this moment en route to the Jovian world. The *big* mission to come is MJS77.

It's not a secret code. MJS77 stands for the twin-robot mission known as Mariner Jupiter/Saturn 1977, and it represents a magnificent new height in interplanetary cooperation for the study of other worlds. Ninety scientists from the United States and four foreign countries—France, Sweden, Germany and the United Kingdom, representing thirty-seven scientific and research institutions—will participate in the most elaborate deep-space mission to date.

The scientists are grouped into eleven areas of investigation, and each investigation area will be represented on the MJS77 Science Steering Group, which will be responsible for the overall science program and will work with the engineers and mission planners to design the spacecraft and the details of the mission.

MJS77 was started in 1971 as a study program under the title of "Outer Planets Missions." In 1977, two Mariner-type robots will be launched by Titan-Centaur boosters to fly by Jupiter, and, using the gravity whip maneuver, be sent on to Saturn. The mission is so planned that the planets will be studied at close range, as well as different moons of Jupiter and Saturn, including—you guessed it—Titan as a high priority target.

Each Mariner will encounter Jupiter after a flight of about twenty months. Three and a half to four years after launch they will reach Saturn. Each Mariner will be fitted out with identical instrumentation, although they may be sent along differing trajectories, expanding the area of exploration.

Before the MJS77 spacecraft are approved in final design, scientists and engineers will wait for the complete results of the Pioneer X and XI flights. Each of the MJS77 robots, however, will carry nuclear generators to power more than a hundred pounds of instruments and television cameras.

And beyond MJS77, now going through final study programs, is an advanced Orbiter robot—a Pioneer modified to go into orbit about Jupiter, first, and then, a second Pioneer to go into orbit about Saturn.

What about the worlds beyond Saturn: Uranus, Neptune and Pluto, drifting through a far domain of stark cold and isolation? A project to send spacecraft to these planets in 1977 and 1979 was well along in development when it ran into the greatest obstacle to such efforts—financial anemia. Everything from Mercury to Saturn met with fiscal approval, but with the waning of full support for scientific programs in space, the ax fell heavily on the outermost worlds of the solar system. No one doubts but that the three last planets outward from the sun roll through space that is appallingly cold and lifeless, and there failed to appear a truly compelling reason to spend money on robot studies of them when other projects were considered far more pressing.

So Uranus, Neptune and Pluto remain in a deep freeze, both in dollars and in actuality. We don't know that much about them, but there's little doubt but that they're frozen across their surfaces, and perhaps through their interior structure. We don't *know* this to be so, but it's certainly going to be a long wait before we find out for sure.

Uranus is a huge globe, nearly 31,000 miles in diameter, or almost four times the diameter of Earth. It's also 14.5 times as massive as our world, but with a density only 1.56 that of water the surface gravity is just about the same we're used to here at home. Uranus swings in its orbit 900 million miles beyond Saturn and is accompanied by five moons. It's a bizarre globe that takes eighty-four years to complete a single orbit about the sun, while it seems to have fallen on its side. For Uranus's axis is tilted so severely—98 degrees out of the plane of its orbit—that for years at a time one or the other of its planetary poles points directly at the sun.

Another giant is Neptune, 33,000 miles in diameter, and with two moons—one of them, Tritan, both larger than Earth's moon and even closer to its mother world. Neptune averages 2,817 million miles from the sun and requires 164 years to complete one solar orbit.

And, finally, there's Pluto, the known end of the solar system. No one is sure of the diameter of the last planet. For years scientists believed Pluto was the same size as Earth, but improved telescopes brought them to scale down the globe to a planet about 3,500 miles in diameter—about halfway between Mercury and Mars. Pluto seems (no one can be sure at these distances) to rotate once on its axis between six and seven days, and it is believed to be a heavy rocky-metallic body, rather than the frozen gases of Uranus and Neptune.

From Pluto, the sun is no more than a bright star in the sky— 4.6 billion miles away—a sun so distant that it requires 248 years before Pluto completes a single orbit about that sun. Yet Pluto insists on being a renegade, moving in an orbit so eccentric that at times it edges closer to the sun than the orbital path of Neptune, leading some astronomers to believe that the ninth planet is really a moon of Neptune that escaped the gravitational chains of what may have been its mother world.

Someday we'll find out.

Wait—there's more. The strange and largely unpredictable nomads of the solar system. The comets, perhaps more than one hundred million of them that are captives of the sun, locked into enormous sweeping orbits that require thousands of years to complete a single revolution of our star.

Astronomers are fascinated by comets, and frustrated as well—because they have managed to chart the orbital paths of only about a thousand of these space gypsies that travel many billions of miles from the sun. There are only a hundred comets, or less, that make a complete orbit of the sun over a period of a hundred years. Forty or fifty comets have been charted to have orbital periods between a hundred and one thousand years.

There's nothing else in our solar system even remotely similar to a comet; the best description might well be a "dirty snowball." There's really no such creature as an "average comet," but a good description, which would apply to many of them, would be to picture a swarm of objects several miles in diameter and made up of dust, pebbles and rocks of all different sizes and shapes, the whole mess "glued together" in a huge glob of frozen ammonia and methane.

When a comet reaches to within 300 million miles of the sun it goes through a magic transformation. It seems to come alive through the radiated energy of the sun striking the nucleus and the sheath. As the comet nears the sun and the solar energy increases, the frozen matter of the nucleus (also called the core) begins to "come alive." Finally it seethes with violent motion. The increasing heat causes explosions that blast dust, debris and gases in every direction. By the time the comet is within 100 million miles of the sun the sheath, the huge tail, is fully established. Solar radiations and especially the solar wind continue to exert mounting pressure on the gases and dust escaping from the nucleus, and we see the comet's tail that can extend as far from the comet as 200 million miles!

Despite their incredible size the comets are nebulous, almost gossamer in their make-up, and their mass is extremely low.

In 1910 the Earth passed directly through the tail of Halley's Comet—a glowing streamer so pale and thin the stars could be seen clearly through the ghostly, illuminated veil. There are times, however, when comets pass near the Earth and much of the rocky debris from the comet is drawn in by Earth's gravity. When that happens the night skies are filled with the dazzling streaks of meteors burning fiercely in our atmosphere.

Halley's Comet didn't need the intense fireworks of meteor trails. It was bright enough in a world in which jet flight, men on the moon, hydrogen bombs, television, bionics medicine and many other miracles were still far in the distant future. Halley's Comet produced a great emotional aura of fear and dread and superstition, as well as delight and awe, and for the next twenty years following its appearance it was perhaps the main topic of conversation for millions of people.

By January 1974 the next great celestial event of this nature will be receding in the night skies. The comet known as Kohoutek graced the skies like a renewed Star of Bethlehem for Christmas 1973.

In early August 1973 I watched the SL-4 Saturn IB booster for the third manned crew of Skylab roll out to its launch pad. We were then in a Rescue Mode, when the Skylab program was in serious trouble. When that problem passed the launch of SL-4 was delayed until November 10, 1973, and we waited for Kohoutek. It was discovered early in 1973 by the German astronomer Dr. Lubos Kohoutek at the Hamburg Observatory, and scientists expected that Kohoutek would be brighter than Venus and visible in daylight.

The details? No one could be sure until the comet started becoming visible in October.

But Kohoutek is only the beginning, and Skylab, which is

STEAMBOAT TIME

planned to study Kohoutek from space for six to eight weeks, is the first stage in a new program of learning extensive details about these strange wanderers of the solar system.

Scientists have an irresistible urge to visit a comet through a Mariner or Pioneer probe. One of the major study programs under way now is to use one of the standby Pioneer X or XI robots on a mission to examine the periodic comet, Forbes, as well as two asteroids, in 1977.

Launch would take place in 1976, but at this moment there's no final decision. One NASA group wants to launch a cometary probe in 1976 to examine the nucleus of the comet Grigg-Skjellerup, with a flyby of the comet in April of 1977. After passing by this comet, the spacecraft would have its orbit adjusted so that in February of 1979 it would sweep by the comet Giacobini-Zinner. The cometary probe would be launched by a Delta booster and would actually be sent into an extreme elliptical orbit of the Earth, then being in position to encounter the comets as they moved by our planet.

And on the horizon is the mission that every astronomer wants urgently to be flown—a launch in 1985 that would take an advanced spacecraft near the nucleus of Halley's Comet.

The wonder of it all . . .

Maybe it's best expressed by a man who's a part of our history. Samuel Clemens, beloved to millions of people as Mark Twain.

Twain once listened to the complaints of an old riverboat pilot, who hated making the changeover from sail to steam. Steeped and locked in his ways, the old man of the river wanted nothing to do with the newfangled steam contraptions, and in a scathing denunciation of "progress" vented his feelings.

"Maybe so," replied Mark Twain, "but when it's steamboat time, *you steam.*"

That's just what we're doing.

Index

ACIC (Aeronautical Chart and Information Center), 116–19
Aeolus as hypothetical Earth, 22–31, 32–38
Aeronautical Chart and Information Center (ACIC), 116–19
Aldebaran, 200
Aminonitriles, 239–40
Apollo VIII, 120–21
Apollo XI, 48, 121–22
Apollo XII, 113, 114, 122–25
Apollo XIII, 125
Apollo XIV, 125–26, 129
Apollo XV, 126, 128
Apollo XVI, 127, 129–30
Apollo XVII, 72–75, 130
Apollo mission, 77, 130–32
Apollo Range Instrumentation Aircraft (ARIA), 158
Arecibo Ionospheric Laboratory, Puerto Rico, 140, 203
ARIA (Apollo Range Instrumentation Aircraft), 158
Armstrong, Neil, 122
Asteroid Belt, 228–30

Bruno, Giordano, 54

Callisto, 237
Cape Canaveral, Florida, 62–63, 77–78
Cayley Formation, 129–30
Cernan, Gene, 130, 133–34
Challenger, 130
Challenger Deep, Marianas Islnds, 42, 43, 166
Chryse, 215, 216
Comets, 243–45

Copernicus, 53–54
Coprates rift valley, Mars, 193, 194, 196
Cro-Magnon, 47, 48, 58
Cydonia, 215, 216

Deimos, 189–92
Descartes Formation, 129–30
Diurnal libration, 118

Eagle II, 148–57. *See also* Venus
Earth, 39–44, 45–58; as part of celestial family, 89–107
Earth science, 128–32
Einstein, Albert, 57
Epsilon Aurigae, 200
Europa, 225–226, 237
Exploration of the moon, 108–32

FDAI (Flight Director Attitude Indicator), 94–95, 105–7, 146–47, 150
Firsoff, V. A., 112, 116
Flight Director Attitude Indicator (FDAI), 94–95, 105–7, 146–47, 150
Forbes Comet, 245

Gagarin, Yuri, 46
Ganymede, 226, 237
Giacobini-Zinner, 245
Glenn, John, 41, 68
Gold, Thomas, 111–12, 203–4
Goldstein, Richard A., 170
Goldstone Tracking Station, Mojave Desert, California, 169
Great Equatorial Current, Jupiter, 224
Great Meteor Crater, Arizona, 230
Great Red Spot, Jupiter, 221, 223, 224
Grigg-Skjellerup, 245

247

INDEX

Hadley Plain, 126
Hadley Rille, 126–27
Halley's Comet, 244, 245

Imbrium Crater, 131
Io, 237

Jet Propulsion Laboratory (JPL), 206
Jupiter, 201, 218–27, 230–33; decametric signals from, 225; temperature variations on, 225; zones and belts on, 223–24

Kitty Hawk, 48, 58
Kohoutek, 244–45
Kordylewski, K., 40

L. Gordon Cooper, 87, 88, 94, 106–7, 134–36, 146, 147, 229
Libration of latitude, 117–18
Longitudinal libration, 118
Lunar mascons, 128–29. *See also* Exploration of the moon
Lunar revolution, 110–11. *See also* Exploration of the moon
Lunar rotation, 110–11. *See also* Exploration of the moon

Mariner I, 160
Mariner II, 142–44, 162–63
Mariner III, 171
Mariner IV, 170, 171, 179–80, 184, 185, 187
Mariner V, 163, 164–65, 167
Mariner VI, 171, 174, 180, 184, 186, 187, 195
Mariner VII, 171, 174, 180, 184, 186, 187, 195
Mariner VIII, 171–72
Mariner IX, 171–75, 180, 184, 187, 189, 192–95, 197–99
Mariner X, 205–11
Mariner pictures, making of, 174–80
Mars, 181–99; major geological provinces of, 194; Russian space flights to, 171–74, 210–15; space interference on way to, 177. *See also* Mariner IV; Mariner VI; Mariner VII; Mariner IX
Mercury, 200–9

MJS77 Science Steering Group, 240–41
Mount Everest, 42, 166

Neptune, 241–43
Neutrino, 55–57
Nix Olympica, 183–84, 194
Noctis Lacus, Mars, 196

Owen, Tobias, 219

Palus Putredinis (Marsh of Decay), 126
Parker, Bruce C., 232
Patrick Air Force Base, Florida, 158–60
Perturbations, 118
Phobos, 189–92
Phoenicis Lacus, Mars, 195–96
Pioneer probes and nuclear energy power, 226–27
Plane of the ecliptic, 104, 105
Pluto, 241–43
Precession libration, 118
Project Viking, 215–17

Radiation belts. *See* Earth; Jupiter
Radioisotope Thermoelectric Generator (RTG), 226

Sagan, Carl A., 170
Saturn, 234–41
Scientific viewpoint, 49–53
Sea of Tranquillity, 58, 125
Shaw, George Bernard, 54
Shklovsky, I. S., 191
Skylab mission, 63, 83, 244, 245
South Tropical Circulating Current, Jupiter, 224
Space flight, beginnings of, 59–88
Strong, John, 144
Sun and the solar system, 200–2

Titan, 235, 237–40
Tithonius Lacus, Mars, 196
Twain, Mark, 245
Tyuratam-Baikonur space center, USSR, 165

Uranus, 241–42
Urey, Harold C., 113–14, 123, 124

Variable Perspective Projector, 117

INDEX

Venus, 146–57; Pioneer spacecraft missions to, 209, 211; Russian probes of, 161–69. *See also Eagle II*; Mariner II
Venus Orbiting Imaging Radar (VOIR), 209

Von Braun, Wernher, 63

Whipple, Fred L., 112, 113, 122

Zodiac, 101